How We Get Mendel Wrong, and Why It Matters

This book illustrates that the stereotypical representations of Gregor Mendel and his work misrepresent his findings and their historical context. The author sets the historical record straight and provides scientists with a reference guide to the respective scholarship in the early history of genetics. The overarching argument is twofold: on the one hand, that we had better avoid naive hero-worshipping and understand each historical figure, Mendel in particular, by placing them in the actual sociocultural context in which they lived and worked; on the other hand, that we had better refrain from teaching in schools the naive Mendelian genetics that provided the presumed "scientific" basis for eugenics.

Key Features

- Corrects the distorting stereotypical representations of Mendelian genetics and provides an authentic picture of how science is done, focusing on Gregor Mendel and his actual contributions to science
- Explains how the oversimplifications of Mendelian genetics were exploited by ideologues to provide the presumed "scientific" basis for eugenics
- Proposes a shift in school education from teaching how the science of genetics is done using model systems to teaching the complexities of development through which heredity is materialized

How We Get Mendel Wrong, and Why It Matters

Challenging the Narrative of Mendelian Genetics

Kostas Kampourakis

CRC Press
Taylor & Francis Group
Boca Raton London New York

CRC Press is an imprint of the
Taylor & Francis Group, an **informa** business

Designed cover image: Kostas Kampourakis, *Making Sense of Genes*
© Kostas Kampourakis 2017, published by Cambridge University
Press—(9781107128132)

First edition published 2024
by CRC Press
2385 NW Executive Center Drive, Suite 320, Boca Raton FL 33431

and by CRC Press
4 Park Square, Milton Park, Abingdon, Oxon, OX14 4RN

CRC Press is an imprint of Taylor & Francis Group, LLC

ISBN: 978-1-032-45691-1 (hbk)
ISBN: 978-1-032-45690-4 (pbk)
ISBN: 978-1-032-44906-7 (ebk)

DOI: 10.1201/9781032449067

Contents

PART I   *Anachronistic Mendelism*

PART II   Social Mendelism

Figures

Tables

Preface: "Gregor Mendel, the First Geneticist"

It is a sunny, calm day in late May 2022. With two talks on Gregor Mendel, his work, and its impact already on my agenda, I am well aware of the bicentennial celebrations related to him. Because of this, I am also curious to read anything new mentioning his name that comes in as an email or social media notification. And, indeed, while checking my email journal alerts, I spot an article about Mendel in the prestigious journal *Nature Reviews Genetics*, titled "The Magic and Meaning of Mendel's Miracle". I have to admit that I become a little concerned when I read about "magic" and "miracles" in relation to science and its history. But the title is success-ful in catching my attention. "Time for a short break from work", I think and I begin reading it. At some point, I read the following:

> Mendel's approach, in contrast, led to a completely new method for measuring the behaviour of living organisms and, as a consequence, triggered a revolution in biomed-ical science, albeit only after several decades' delay. His work was not widely dissemi-nated and was too radical and abstract for his contemporaries to grasp. . . . As a result, his work was never appreciated in his lifetime, and his example, therefore, provides a beacon for young people embarking on a scientific career. You must travel with your own moral compass and must believe in yourself and that posterity, not peers, should be your true judge. As Mendel himself so aptly put it when questioned about his neglect: "My time will come".[1]

The author of these words is Kim Ashley Nasmyth FRS, an English geneticist, the Whitley Professor of Biochemistry, and a fellow of Trinity College at the University of Oxford. He describes a scientific hero: a genius who was ahead of his time, who died unappreciated by his scientific peers even though his work initiated a revolu-tion in the life sciences, and whose major contribution was recognized only several decades later. Nasmyth compared Mendel's contribution to that of Newton: "Like Newton, who imagined a magical new force to explain the movement of the planets, Mendel imagined a similarly magical process, namely the reciprocal segregation of character difference determining elements during gamete formation".[2] "How sad" you may think. . . . Why such a genius was not appreciated for his key discovery? Why did it take several decades for scientists to realize the importance of Mendel's work? How could people be so oblivious?

Mendel shares another key feature with Newton: they are common instantiations of the "lonely genius" stereotype—a (white) man working in isolation and discovering the laws of nature. He is also one of the few scientific figures in the life sciences that you likely remember from your school years—perhaps along with Charles Darwin, Louis Pasteur, James Watson, and Francis Crick. I would not be surprised if you still remember the crosses you did at school, which showed how traits are inherited—and based on which your biology teacher might have asked you to explain why two parents with blue eyes cannot have a child with brown eyes (more on this later). Mendel conducted experiments with peas in his monastery during the 19th century, and, according to most biology textbooks, he discovered the laws of heredity and set the foundations of genetics. Here is, for instance, how the prestigious journal *Nature* represents Mendel and his work on its educational pages:

> **The Father of Genetics.** Like many great artists, the work of Gregor Mendel was not appreciated until after his death. He is now called the "Father of Genetics".
>
> **Mendel's brilliance is unrecognized.** On February 8, 1865, Mendel presented his work to the Brunn Society for Natural Science. His paper "Experiments on Plant Hybridization" was published the next year. While his work was appreciated for its thoroughness, no one seemed to grasp its importance. The work was simply too ahead of its time, too contrary to popular beliefs about heredity. "My time will come", Mendel once said, but it was over 30 years before his work was appreciated.[3]

Like Nasmyth, this account of Mendel's work tells us that he was a genius who discovered the secrets of heredity and paved the way for the science of genetics, being belatedly recognized as the "father" of the discipline. Why belatedly? In part, because he worked in isolation and therefore was outside the mainstream scientific communities of his time; in part, because his thinking was ahead of his time and so too difficult for others to understand. Or so the narrative goes. This is the one that I am challenging in the present book.

This is a book about science: its theories and their history. Its aim is to show why these are intertwined and why it is not possible to understand the former without the latter. The scientific discipline under focus is genetics, which studies heredity and spans all the related phenomena from individual development to DNA variation in populations. This science is said to have its origins in the mid-19th century, in the work of the isolated humble genius Mendel. His name is associated with a specific field of study "Mendelian genetics", which is often of interest in many important endeavors of everyday life such as medicine or plant and animal breeding. In 2022, many events around the world celebrated the bicentenary of Mendel's birth repeating the narrative presented earlier again and again. My aim in the present book is to challenge the stereotypical narrative of Mendelian genetics about his life and work.

Before we proceed, I should state clearly what kind of historical book this is so that readers do not bring the wrong expectations to it. Even though this is a book that could be of value to anyone interested in the history of science, especially of the life sciences, it is primarily intended for life scientists, students in the life sciences, and biology teachers (and this is why I assume that readers have some previous basic knowledge about DNA, genes, chromosomes, and cell division).[4] What I am trying to do in the present book is to explain why the actual story of Mendel is more complicated, and also quite different, than what the stereotypical narrative of Mendelian genetics indicates. In fact, I hope that the present book will serve as an antidote to this narrative by explaining why much of it is historically inaccurate, while also setting the historical record straight. This is something I have tried to achieve in several previous writings, starting 20 years ago,[5] but eventually I decided that a book-length account was necessary. A secondary aim of the present book is to serve as an introduction to the scholarship in the history of science, for those who want to read more about the various episodes discussed.

However, this is not a book of the academic history of science. The historical scholarship I draw upon is selective, and far from exhaustive. I am actually aware that the kind of history of science included herein might strike some historians of science as old-fashioned, in being based entirely on published primary and secondary sources but not on any archival research, and in concentrating on scientific concepts in the writings of a handful of (mostly white and male) scientific thinkers, with little information about the wider scientific community they belonged to, let alone the changing, globalizing social world in which that community was embedded, and in which heredity came to matter more and more. Rather, this is an educational book that draws on the history of science in order to make a general point about why non-historians should carefully and thoughtfully study it, rather than rely on myths and falsehoods. I write this book as a biology education specialist long been convinced that perspectives from the history and philosophy of science are indispensable in avoiding unhelpful oversimplifications about biology. My approach to the historical narrative I am challenging is that of someone synthesizing a specialist history of science literature for an audience unfamiliar with its insights. I am therefore using that history to make a hopefully useful, and in my view much-needed, intervention in today's discussions about Mendel and his legacy. It is to that end that I have made the choices I have about what to emphasize and where. All these lead to arguing toward the end of the book that naive Mendelian genetics thinking is not just scientifically and historically inaccurate but also socially problematic and sometimes dangerous (see Chapters 8 and 10).

The narrative that presents Mendel as the founding father of genetics, who understood that inheritance was particulate in nature, who discovered the laws of heredity, who was ignored by his contemporaries, and whose reputation was established posthumously in 1900 with the rediscovery of his pioneering paper, has been critically questioned since at least 1979.[6] The notion that scientific discoveries suddenly emerge, during eureka moments, in the minds of brilliant scientists is a recurrent myth that distorts the fact that science is done by communities, not individuals. There are no fathers (or mothers or any other ancestors for that matter) of disciplines, as everyone who has made any kind of contribution to a scientific field has always had to rely on the work of those who came before them—stand on the shoulders of giants, as the saying goes. Mendel is not the father of genetics, but a brilliant hybridist who developed an experimental approach that was later used in a new science that was called genetics in 1906. If people failed in 1866 to grasp the importance of his work, it was because it was irrelevant to what most naturalists studying heredity were doing at the time (see Chapter 3). As I explain in the present book, Mendel's experimental approach became useful when it was integrated into a new framework that did not exist in his time (see Chapter 4). Mendel's work can be said to be ahead of its time only in hindsight and under an anachronistic reading of history (see Chapter 5).

Paying attention to the details of history matters for various reasons. For instance, in most accounts I have seen, Mendel is described as a monk. But he was not a monk; he was a friar. "OK, but so what?" you may think. Well, this is one of the details that matter. Whereas being a friar is similar to being a monk, in the sense of belonging to a religious order of the Catholic Church and being devoted to a religious life, there is also a very important difference. In contrast to monks who live in secluded groups, friars live and work among other people and participate in societal life. This is of critical importance for realizing that Mendel was not working in isolation but rather carried out his experiments in a specific socioeconomic context with agricultural interests (see Chapter 2). His research was intended to provide answers to particular practical questions about agriculture that were of importance in the place where he lived and worked. Therefore, Mendel was strongly integrated into his society and was thus very much a man of his time. This is the kind of corrective of stereotypes that I provide in the present book, with the aim to help readers not only understand history but also why its careful study matters.

History is important because its study can help us understand how we got where we are now; it is not only about who did what and when but also about understanding the content of science itself. For instance, by studying the history of science we can understand how the concepts we currently use were coined and have evolved ever since. Mendel did not mention anything that might resemble the concept of genes we use today. But even Wilhelm Johannsen, who coined the term "gene" in 1909, was not specific about what a gene is (see Chapter 7). In his Nobel Prize lecture of 1933, the first Nobel Prize in Physiology and Medicine to be awarded for research in genetics, Thomas Hunt Morgan argued that at the level at which research was done, it did not really matter whether genes were real entities or not.[7] Compare this to the stereotypical view that Mendel inferred the presence of genes on the basis of his experimental findings. In hindsight, given our current knowledge, it is indeed easy to read too much into Mendel's paper and conclude that he discovered the laws of heredity that today have his name. But if we manage to read his paper in its original context, we might figure out some of its outstanding achievements due to which it was co-opted as the founding document of genetics after 1900, while avoiding the seduction of anachronism and naive hero-worshipping. I should note at this point that it is not my aim with this book to dethrone Mendel. But if we want to give Mendel credit for anything, we first need to understand what he did and what he did not do. Mendel's paper had no impact before 1900 but became the founding document of genetics after 1900, and this was done for good reasons. Mendel did not have the intention to do genetics, but his brilliant experimental approach was found to be useful when the discipline of genetics emerged.

But when and how was Mendel established as the founder of the discipline? There are at least two crucial points in time. The first was the publication in 1902 of William Bateson's book *Mendel's Principles of Heredity: A Defence*, which introduced Mendel's work to the English-speaking world. But Bateson did more than this. He vehemently defended Mendel's work and principles against the criticisms of Raphael Weldon (see Chapter 6).[8] In his Preface, Bateson wrote: "two years ago it was suddenly discovered that an unknown man, Gregor Johann Mendel, had, alone, and unheeded, broken off from the rest—in the moment that Darwin was at work—and cut a way through".[9] "Alone, and unheeded . . . cut a way through"—here is perhaps the first "lonely genius" account of Mendel's work. Bateson also expressed his confidence in the importance of Mendel's work:

I venture to express the conviction, that if the facts now before us are carefully studied, it will become evident that the experimental study of heredity, pursued on the lines Mendel has made possible, is second to no branch of science in the certainty and magnitude of the results it offers.[10]

Thus, Bateson affirmed the importance of Mendel's work and findings in 1902. He concluded his defense of Mendel's principles with a somewhat prophetic statement:

In these pages I have only touched the edge of that new country which is stretching out before us, whence in ten years' time we shall look back on the present days of our captivity. Soon every science that deals with animals and plants will be teeming with discovery, made possible by Mendel's work. The breeder, whether of plants or of animals, no longer trudging in the old paths of tradition, will be second only to the chemist in resource and in foresight. Each conception of life in which heredity bears a part—and which of them is exempt?—must change before the coming rush of facts.[11]

The second crucial year is 1965. This was the centenary of the presentation of Mendel's paper, which was celebrated in various ways. One was the publication of book-length histories of genetics written by prominent geneticists, such as Alfred Henry Sturtevant and Leslie Clarence Dunn. Sturtevant was a doctoral student and subsequently a close collaborator of Thomas Hunt Morgan. Among his many contributions to genetics, he is remembered for being the one to suggest that the distance between two genes on a linear map can be inferred from the frequency of crossing over between them and devised a way for mapping genes on chromosomes (see Chapter 7). The book *A History of Genetics* was his last major work.[12] Dunn was one of the most productive researchers in genetics and the author of books that influenced a whole generation of geneticists. He was a doctoral student of William E. Castle and did exceptional research in the genetics of mice and poultry. He was also the co-author, with Edmund W. Sinnott, of the book *Principles of Genetics*, which was published in 1925 and soon became one of the most widely used books in genetics.[13] A third, perhaps less well-known, history of genetics of that time was written by agronomist and plant breeder Hans Stubbe; it was published in German in 1965 and was translated into English in 1972.[14]

All these authors of the 1965 histories of genetics blamed the Swiss botanist Carl Wilhelm von Nägeli, with whom Mendel had a long correspondence from 1866 to 1873,[15] for the neglect of Mendel's work. Sturtevant wrote about Nägeli: "He completely failed to appreciate Mendel's work and made some rather pointless criticisms of it in his reply to Mendel's letter. He did not refer to it in his publications". Sturtevant went on to argue that Mendel had so much confirmatory evidence that had it been published, it would have been unlikely for it to have been "so completely ignored . . . even though it was not enough to convince Nägeli".[16] Dunn agreed that Nägeli, "one of the foremost important botanists and hybridizers of his time, failed to mention Mendel or his theory either in papers published after the extensive correspondence with Mendel or in his major work on heredity". Nägeli's statements in his correspondence with Mendel, as well as the "absence of comment on Mendel's theory" in his subsequent work, indicates that Nägeli "had missed the essential feature of Mendel's demonstration and did not recognize the significance of Mendel's work".

Dunn even described Nägeli's advice to Mendel as that of "a highly placed professional patronizing an amateur".[17] In a similar spirit, Stubbe wrote that "it is quite clear from his critical remarks about Mendel's essay that Nägeli did no[t] appreciate its significance". Even though Nägeli cited Mendel's work, according to Stubbe, "the fact that Nägeli merely mentions the essay [Mendel's] shows even more clearly that he did not understand the importance of that work".[18]

In his book, Dunn pointed out that historians of science had paid little attention to the history of genetics. But it seems that he did not know, and perhaps could not have known, that historian Robert Olby was at the same time writing such a book. Published in 1966, Olby's *Origins of Mendelism* is still today an authoritative account of the development of genetics (a second, crucially enriched edition was published in 1985). According to Olby, Nägeli was interested in Mendel's results. However, he could not believe that Mendel was right in his conclusions, because Nägeli was quite sure that all the offspring of hybrids were variable, and that "true-breeding" hybrids, that is, ones that always passed down to their descendants the same phenotypic traits, could not exist. In other words, Nägeli thought that even if Mendel had found so-called constant forms, that is, forms resembling the parental ones after crossing hybrids, there would still be variation in subsequent generations.[19] Olby thus suggested that Nägeli disagreed with Mendel, rather than that he failed to appreciate the significance of his findings.

In Part I of the present book, titled "Anachronistic Mendelism", I try to dismantle the anachronistic narrative according to which Mendel supposedly discovered the laws of heredity in 1865. Anachronism is the transfer of a cultural element, in this case, a piece of scientific knowledge that characterizes a particular historical period, into a narrative that refers to another historical period. I present Mendel's work in the context of his own time, and I distinguish that from the co-option of his work 34 years after its publication. Part II is titled "Social Mendelism", a term I have borrowed from historian Amir Teicher, who has defined it as "the extension of Mendelian thinking to the human domain".[20] Eugenics was neither primarily nor mostly an instance of Social Darwinism, that is, of the application of Darwin's biological theories to the social and cultural realm. In Part II, I show how after Mendel's paper became the foundational document of the new science of genetics, oversimplistic accounts of heredity initially developed for domesticated plants and animals were applied to human heredity and supported eugenic policies with very harmful outcomes.

Let our foray into the history of genetics begin!

NOTES

1 Nasmyth (2022), p. 448.
2 Nasmyth (2022), p. 452.
3 www.nature.com/scitable/topicpage/gregor-mendel-a-private-scientist-6618227/ (accessed June 18, 2022).
4 For those readers who do not, Kampourakis (2021a) provides all the necessary background knowledge, and more.

5 Kampourakis and Roumeliotou (2004); Kampourakis (2013), (2015), (2017), (2021a), (2021b).

6 Olby (1979); Brannigan (1979).

7 Thomas H. Morgan—Nobel Lecture. NobelPrize.org. Nobel Prize Outreach AB 2023. Thu. 22 Jun 2023. www.nobelprize.org/prizes/medicine/1933/morgan/lecture/

8 Radick (2023).

9 Bateson (1902). p. v.

10 Bateson (1902) pp. ix–x.

11 Bateson (1902), p. 208.

12 Lewis (1978).

13 Dobzhansky (1978).

14 Stubbe (1972).

15 Mendel (1950). In 1950, to celebrate the 50th anniversary of the rediscovery of Mendel's work, the Genetics Society of America published a special supplement, containing Mendel's letters to Carl Nägeli as well as translations of the original papers by Carl Correns, Hugo de Vries, and, Erich von Tschermak.

16 Quotations from Sturtevant (1965), pp. xi, 11, 12.

17 Quotations from Dunn (1965), pp. 16–17, 19.

18 Quotations from Stubbe (1972), pp. 156–157, 160.

19 Olby (1966), p. 118.

20 Quotations from Teicher (2020), p. 5.

About the Author

Kostas Kampourakis is the author and editor of several books about evolution, genetics, philosophy, and the history of science. He teaches biology and science education courses at the University of Geneva, Switzerland. He is the co-editor of *Teaching Biology in Schools* and *What Is Scientific Knowledge*, both published by Routledge.

Acknowledgments

Writing a book is a solitary endeavor, with a huge debt to those scholars who came before, as well as those in the present who devote significant amounts of time to reading one another's work and providing constructive feedback. For their valuable comments and suggestions, I am therefore indebted to the following scholars (in alphabetical order): Nathaniel Comfort, David Curtis, Oren Harman, Jon Marks, Staffan Müller-Wille, Diane Paul, Erik Peterson, Greg Radick, Andrew Reynolds, George Davey Smith, James Tabery, and Amir Teicher. But I owe even more to all these people for various reasons. Greg Radick has done complementary work to my own, and we have had very fruitful scholarly interactions throughout the years. His recent book *Disputed Inheritance* is the apogee of his scholarship on all things Mendelian, and the book you must definitely read after the present one. Erik Peterson and I have also had important scholarly interactions over the years, beginning with a 2015 Mendel special issue in the journal *Science & Education*, where we tried to correct distortions. Staffan Müller-Wille's work has been extremely influential for me, as has been his translation (with Kersten Hall) of Mendel's paper. Nathaniel Comfort's writings are a constant light in the darkness of genetic determinism. Diane Paul is one of the best eugenics scholars in the world, and her writings have clarified all aspects of this story. Amir Teicher's book *Social Mendelism* was a revelation for me and filled in a gap that had been troubling me for about 20 years. I learned about Lancelot Hogben from Jim Tabery's book *Beyond Versus*, as well as from his other writings. Jon Marks has written about issues similar to those raised in the present book a long time ago, and his writings have always been an inspiration. Andrew Reynolds has written excellent works on the use of metaphors in the history of science that made me interested in this topic too. Oren Harman's work is the one that fills in many important gaps in scholarly work, even where you had not even realized there was one. Finally, George Davey Smith invited me to contribute to a fascinating Mendel conference he organized in Bristol, in July 2022, which motivated me to write this book. Further motivation came from discussions with David Curtis after that conference.

I am also very grateful to many very kind and helpful people who provided me with many of the photos in the present book: Amir Teicher and Marsha Richmond for the photos from their previous works; Kate West and Sarah Wilmot for the photos from John Innes Archives; Ulrike Denk for the images from the archives of the University of Vienna; Penny Neder-Muro for the photos from the Caltech Archives & Special Collections; Stefan Sienell for the photo from the archives of the Austrian Academy of Sciences; Michael Miller for the photos from the archives of the American Philosophical Society; Will Rossiter for the photo from the University of Adelaide Library, Rare Books and Manuscripts; Annie Moots for the photo from Pickler Memorial Library, Truman State University; Dan Mitchell and Rafa Siodor for the photos from UCL Library Services—Special Collections; Cambridge University Press for the permission to use two images from two previous books of mine. I am also very grateful to Lynn Chiu and Barbara Fischer, organizers of the 2022 KLI Mendel symposium who, along with Blanka Křížová, Director of the Mendel Museum in Brno, arranged a very informative tour at the Mendel Museum and the monastery, during which I took some photos that I have included in the present book. Finally, I am indebted to the people at the Marine Biological Laboratory Archives (https://history.archives.mbl.edu) and the U.S. Holocaust Museum (www.ushmm.org) who have made photos at high resolution and at no cost available for anyone who might want to use them. In all cases, I have aimed to include photos of people around the same time as the events I am describing or the writings I am presenting, and I hope I have achieved that.

Speaking of archives, it is amazing how many primary sources are available online today. It is therefore worth acknowledging some of the places that have made available most of the primary sources I drew upon for my work in this book, such as the Wellcome Collection, the Electronic Scholarly Publishing Project, the Biodiversity Heritage Library, the Darwin Online project, the Darwin Correspondence Project, the galton.org website, and the Internet Archive among others.

Last but not least, I owe many thanks to Chuck Crumly for supporting this book and to Kara Roberts for her extremely precious help from the proposal stage to publication.

As ever, I am grateful to my family for their love and support. They still wonder how writing books can count as a hobby, but they do accept that it might, as they see the pleasure it gives me. I hope that reading this book will give you pleasure too, by finding it both interesting and useful.

Gregor Johann Mendel (1822–1884).
(Courtesy of Professor William Bateson,
London.)

FIGURE 0.1 Credit: Portrait of Gregor Johann Mendel, Garrison. Wellcome Collection.
Attribution 4.0 International (CC BY 4.0).

Part I

Anachronistic Mendelism

1 Mendel Was not a Geneticist ahead of His Time

THE STEREOTYPICAL NARRATIVE ABOUT MENDEL'S WORK AND FINDINGS

The stereotypical narrative about Mendel's work, the important discoveries he made, and how they fell into oblivion for 34 years is a story one often finds in biology school textbooks. Mendel is commonly portrayed in textbooks and elsewhere as working alone in his garden, where he discovered the laws of heredity (Figure 1.1). Along with Charles Darwin, he is one of the historical figures most likely to encounter in any biology textbook. Crucially, Mendel is often presented as the most heroic figure of all. In contrast to Darwin, who achieved fame during his lifetime and is currently considered as the one who set the foundations of evolutionary biology with his book *On the Origin of Species*, published in 1859, Mendel died in oblivion and only retrospectively, and several years after he died, was he resurrected as the founding father of genetics.

Let us consider some examples of textbook accounts about Mendel being the father of genetics:

- The study of genetics, which is the science of heredity, began with Mendel, who is regarded as the father of genetics.[1]
- The modern science of genetics was started by a monk named Gregor Mendel.[2]
- The groundwork for much of our understanding of genetics was established in the middle of the 1800s by an Austrian monk named Gregor Mendel.[3]
- Modern genetics had its genesis in an abbey garden, where a monk named Gregor Mendel documented a particulate mechanism for inheritance.[4]
- Gregor Mendel is considered to be the founding father of genetics.[5]

As well as about what exactly he "discovered":

- For the next 35 years, his [Mendel's] paper was effectively ignored yet, as scientists later discovered, it contained the entire basis of modern genetics.[6]
- In his 1866 paper "Experiments in Plant Hybridisation" the Moravian monk Gregor Mendel laid down the basis for the entire subject of genetics by postulating for the first time the existence of the discrete hereditary determinants we now call genes.[7]

DOI: 10.1201/9781032449067-2

3

FIGURE 1.1   The stereotypical image of Gregor (Johann) Mendel (1822–1884), working alone in his garden. This is the single best illustration of the "lonely genius" stereotype of a scientist. From Kostas Kampourakis, *Making Sense of Genes* © Kostas Kampourakis 2017, published by Cambridge University Press, figure 1.4, reproduced with permission.

- At the St. Thomas monastery in the mid-1800s, Mendel carried out both his monastic duties and a groundbreaking series of experiments on inheritance in the common edible pea. It was not until 1900 that three biologists—Carl Correns, Hugo de Vries, and Erich Tschermak—working independently and knowing nothing of Mendel's work, rediscovered the principles of inheritance.[8]
- This segregation of alternative forms of a character, or trait, provided the clue that led Gregor Mendel to his understanding of the nature of heredity.[9]
- Darwin never learned of Mendel's work because it went unrecognized until 1900. . . . At the time Mendel began his study of heredity, the blending concept of inheritance was popular. Mendel carefully designed his experiments and gathered mathematical data to arrive at a particulate theory of inheritance rather than a blending theory of inheritance.[10]

The common assumption in all textbooks is obvious: Mendel set the foundations of genetics, studying heredity; some textbooks even claim that he discovered that heredity is particulate in nature and that the factors controlling it are those we now call genes. Let us see in more detail what exactly he is said to have "discovered".

Mendel is credited with the discovery that traits are controlled by hereditary factors, the inheritance of which follows two laws: the law of segregation and the law of independent assortment. In the first case, when two plants that differ in one trait, for instance, their seeds are either round or wrinkled, are crossed, their offspring (this is called filial generation 1, or F_1) resemble one of the two parents (in this case, they have round seeds). In generation 2 (the offspring of the offspring, or filial generation 2, or simply F_2), the ratio between the round and the wrinkled trait is 3:1 (Figure 1.2). But Mendel is not only presented as just making these observations; he is also presented as having provided an explanation for them. According to textbook accounts, round shape is controlled by the dominant factor R, whereas wrinkled shape is controlled by the recessive factor r. That R is dominant entails that when an individual has an R and an r factor in their genotype (this is how we describe what alleles—the different versions of genes—an individual has), it is R that dominates r as it imposes its respective phenotype (round) over that of r (wrinkled). In short, an individual with genotype Rr (described as heterozygous because it has two different alleles) has the same phenotype—round seeds—as individuals with genotype RR (described as homozygous because it has the same allele twice; this is also how an individual rr that has wrinkled seeds is described). Mendel's achievement, according to textbook accounts, was that from the respective ratios in generations F_1 and F_2, he managed to infer that the factors (R/r) controlling the different phenotypes (round/wrinkled) are separated (segregated) during fertilization and recombined in the offspring. This is described as Mendel's law of segregation.

Mendel also studied the simultaneous inheritance of two traits, for instance, the shape of the seeds and their color. When he crossed plants with yellow/round seeds and plants with green/wrinkled seeds, in F_1 all plants had yellow/round

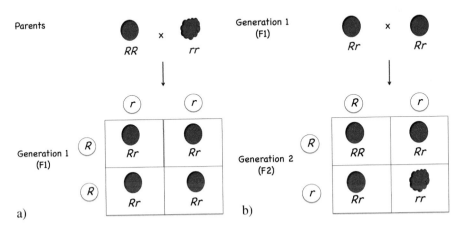

FIGURE 1.2 A cross between two plants that differ in the shape of seeds (round or wrinkled). Plants with round seeds have factors *RR* or *Rr*, whereas plants with wrinkled seeds have factors *rr*. The "wrinkled" character "disappears" in F1 (a) and "reappears" in F2 (b).

seeds. However, when those plants were left to self-fertilize or were crossed with one another, a constant ratio of 9 yellow/round:3 yellow/wrinkled:3 green/round:1 green/wrinkled emerged in F_2. This is actually the result of the combination of the probabilities to have all possible combinations of two characters, for instance, one for the shape and another for the color. So, what happens is this: the crossing of two individuals, one with round seeds and one with wrinkled seeds, gives a ratio of 3 round:1 wrinkled in F_2 (Figure 1.2). Similarly, the crossing of two individuals, one with yellow seeds and one with green seeds, gives a ratio of 3 yellow:1 green in F_2. Therefore, their combination gives the following ratio:

3 yellow × 3 round	→	9 yellow/round
3 yellow × 1 wrinkled	→	3 yellow/wrinkled
1 green × 3 round	→	3 green/round
1 green × 1 wrinkled	→	1 green/wrinkled

Mendel is said to have inferred from these results that the factors (R/r and Y/y) controlling the different traits (seed shape and seed color, respectively) were assorted independently during fertilization. As a result, all possible combinations were obtained (yellow/round, yellow/wrinkled, green/round, green/wrinkled), and this is why the ratio 9:3:3:1 is observed in F_2 (see Figure 1.3). This is described as Mendel's law of independent assortment.

These are the results that Nägeli supposedly failed to understand, as we saw in the Preface, according to Dunn, Sturtevant, Stubbe, but also several others. For instance, Edward Murray East, one of the founding researchers of maize genetics, wrote in 1923 about Nägeli:

> Here was a man to whom Mendel had written in detail about his work during the years between 1866 and 1873, a man who had contributed notable papers to botany

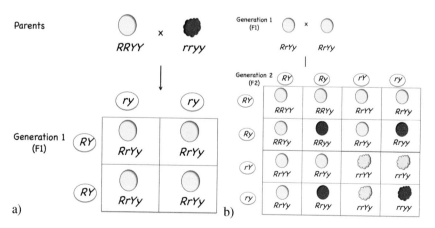

FIGURE 1.3 (a) A cross between two plants that differ in the shape of seeds (round/wrinkled) and the color of seeds (yellow/green). (b) In F2 we find the characteristic ratio 9:3:3:1 (green seeds appear here as having a darker color than yellow seeds).

on subjects ranging from the form of the starch grain to hybridization, a man who discoursed at such length on chemistry and physics that one might suppose him to have had the greatest sympathy for the highest type of useful quantitative work; but he devotes absolutely no time or energy to discussing the one paper which might have shown him a way out of the wilderness in which he found himself.[11]

It is now time to consider what exactly Mendel wrote in his famous paper.

WHAT MENDEL ACTUALLY WROTE IN "EXPERIMENTS ON PLANT HYBRIDS"

Johann Mendel was born in Moravia in 1822, which was then a province of the Austrian Empire (today it is part of the Czech Republic). Growing up he had to deal with financial hardships, and even though he had intended to study at the university, by the age of 16 his parents were unable to support him financially. When in 1840 he enrolled at the Philosophy Institute at Olomouc, Czech Republic, in order to complete a two-year course that was a prerequisite for enrolling in university studies, his younger sister had to renounce part of her dowry in order for Mendel to be able to finance his studies. Thus, when at the recommendation of Professor Franz, who taught Mendel physics, Mendel was given the opportunity to enter the Augustinian monastery of St. Thomas in Brno, he took the offer in order to stop struggling for his survival. Franz knew Cyrill Napp, the Abbott of the monastery, who accepted Mendel there in 1843, thereafter assuming the name Gregor (Figure 1.4). After he

FIGURE 1.4 Gregor Mendel (standing, fourth from the right) and Cyrill Napp (sitting, center). Photo taken by the author at the Mendel Museum in Brno.

completed his theological studies, Mendel taught for some time at a Gymnasium at Znojmo; however, in 1850 he failed to pass his teacher examination at the University of Vienna, having done well in physics but not in zoology. Nevertheless, he was willing to continue studying, and the monastery in which he lived was an environment favorable to further studies. Indeed, Napp supported Mendel's studies in Vienna from 1851 to 1853.[12]

Hybridization studies were considered of interest in the socioeconomic context of Brno (see Chapter 3). Mendel began preparing his hybridization experiments in 1854. During two years, he studied 34 distinct varieties of the edible pea (*Pisum sativum*), which differed in 15 traits. Among those, he selected 22 varieties that did not vary over two years, that is, when they self-reproduced, they always produced plants with the same traits. Mendel then performed crosses between different varieties, focusing on seven (binary) characters:

- the shape of the ripe seeds (round or wrinkled)
- the color of the ripe seeds (yellow or green)
- the color of the seed coat (white or gray-brown)
- the shape of the ripe pod (smoothly arched or deeply ridged between seeds)
- the color of the unripe pod (light to dark green or bright yellow)
- the position of flowers (axillary or terminal)
- the length of the stem (1.9–2.2 m or 0.24–0.46 m)

It is important to note that Mendel admitted in his paper the existence of variation in traits that did not allow for a "sharp separation". He considered such traits as "unsuitable" for his experiments and selected only those for which a clear distinction between two types was possible.[13]

Mendel began his experiments in 1856 and completed them in 1863. A notable feature of his work was the mathematical analyses he did, having crossed a very large number of plants (28,000 according to one estimation, of which he carefully studied 12,835). Mendel wrote and presented his results in the form of two presentations during the meetings of the Brno Natural Science Society on February 8 and March 8, 1865. He later had his written account submitted for publication, which he titled "Versuche über Pflanzen-Hybriden" ("Experiments on Plant Hybrids") and which appeared in volume IV of the journal *Verhandlungen des naturforschenden Vereins zu Brünn* in 1866 (Figure 1.5).[14]

To better understand what exactly Mendel did, it is useful to consider what exactly he wrote in that famous paper.[15] Mendel began by stating that the aim of his experimental work was "to follow up the development of hybrids in their descendants", which had been inspired by "the striking regularity with which the same hybrid forms always recurred whenever fertilisation happened between like species". He pointed out that up to that time no one had succeeded "in establishing a generally valid law for the formation and development of hybrids" and that among the various experiments that had been conducted until his time, none was as extensive as necessary in order to study large numbers of hybrid offspring. In short, Mendel made clear right from the start that he was interested in the formation and development of hybrids. He also suggested that the fact that there was until then no widely accepted

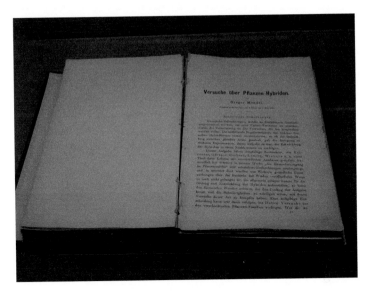

FIGURE 1.5   Mendel's paper in the Brno Natural History Society journal. Photo taken by the author at the Mendel Museum in Brno.

law for this process was due to the inadequate experimental work done, stating right afterward that this was the main contribution that his experimental work made. This is also evident from the fact that he mentioned the names of some important plant hybridists, in particular Kölreuter, Gärtner, Herbert, Lecocq, and Wichura, even though in his paper he directly refers only to Gärtner's work (see Chapter 2).[16]

Mendel explicitly stated two more times in his paper that his aim was the formulation of a law driving the formation of hybrids: "To observe these changes for two differing traits respectively, and to investigate the law according to which they occur in the successive generations, was the task of the experiment":[17]

> In the experiments just reviewed, plants were used that were different in one essential trait only. The next task consisted in investigating whether the developmental law that was thus found is also valid for two differing traits respectively in those cases where several different characteristics are united through fertilization in the hybrids.[18]

The phrase "two differing traits respectively" refers to trait pairs, such as round/ wrinkled or yellow/green seeds. Mendel thus very clearly stated that he was interested in figuring out the law according to which the transmission of traits occurred in the descendants of hybrids. As historians Müller-Wille and Hall have argued, regardless of whatever Mendel meant, it is important that he referred to one law, not more than one as later geneticists, and textbooks still today, have attributed to him. This law seems to be clearly described toward the end of his paper:

> The law of combination of differing traits, according to which the development of hybrids occurs, thus finds its rationale and explanation in the proven proposition that

the hybrids produce germ- and pollen-cells that correspond in equal quantities to all the constant forms that emerge from the combination of the traits that were united by fertilization.[19]

In short, Mendel was explicit in his paper about what he wanted to do and what he actually did: figure out the law that drives the development of hybrids.

After describing the experimental procedure that he followed and the traits he studied, Mendel turned to what he had found. "Each of the seven hybrid traits resembles one of the two parental traits either so completely that the other one seems to escape observation, or is so similar to the same that a reliable distinction cannot be made". What Mendel described here is, simply put, that the traits exhibited by the plants in the first generation (these were the hybrids) were similar to those of one of the parental plants. It is important to note that the term "hybrid traits" should be interpreted as referring to traits that themselves were hybrid, not to the traits of hybrids. In other words, Mendel considered that the traits of the parental plants were combined in the hybrids, which in turn exhibited traits that were themselves hybrid. However, these hybrid traits were not intermediate between the parental traits but resembled only one of them. Mendel then explained how these were described and distinguished: "In the following review, those traits that pass into the hybrid conjunction entirely or almost unchanged, hence themselves representing the hybrid traits, will be denoted as *dominating*, and those which become latent in the conjunction as *recessive*". This is a key sentence that has often been misinterpreted. It is the traits that are dominating (hereafter dominant)[20] or recessive; the former were those that were passed unchanged to the hybrids, whereas the latter were those that remained invisible in their conjunction with the former. Mendel explained that he chose the term "recessive" because these traits might disappear completely in the hybrids and then reappear in their offspring.[21] In short, a hybrid trait actually comprised two traits of which only one was visible (which was described as dominant), whereas the other was not (which was described as recessive).

Then Mendel considered what happened in the second generation, that is, the offspring of the hybrids.

> In this generation, alongside the dominating traits the recessive traits reappear as well in their full peculiarity, and they do this in a decisively pronounced average proportion of 3:1, such that among 4 plants each from this generation 3 plants receive the dominating and one plant the recessive characteristic. This applies without exception to all the traits that were included in the experiments.[22]

Mendel then suggested that as all hybrids were similar, the results from reciprocal crosses could be considered as results from a single hybridization experiment. Reciprocal crosses are those in which males of kind A are crossed with females of kind B, while females of kind A are also crossed with males of kind B. As peas are monoclinous plants, that is, they have stamens and pistils in the same flower, there are no "male" and "female" plants.[23] Simply put, as there was no difference among the results of crosses, say A × B, depending on whether pollen A was transferred to pistil B or vice versa, the total results could be considered altogether. Mendel thus presented the total results:

1st trial: Seed shape. From 253 hybrids 7324 seeds were obtained in the second year of the experiment. Among these, 5474 were round or roundish and 1850 were angular and wrinkled. This results in a proportion of 2.96:1.

2nd trial: Colouration of the albumen. 258 plants gave 8023 seeds, [of which] 6022 [were] yellow and 2001 were green; therefore the former stand in a proportion of 3.01:1 to the latter.[24]

Mendel continued the presentation of these results, writing: "Once the results of all the experiments are drawn together, an average proportion of 2.98:1, or 3:1, emerges between the quantities of forms with the dominating and the recessive trait". He then concluded that section with the following statement:

> The dominating trait can have a double meaning here, namely that of parental characteristic or that of hybrid trait. In which of the two meanings it appears in each individual case can only be decided by the next generation. As a parental trait, the same must devolve without alteration to all descendants, as a hybrid trait on the other hand, it must observe the same behaviour as in the first generation.

According to Müller-Wille and Hall, the "meaning" of traits depended not only on their manifest qualities but also on what happened to them in succeeding generations. The dominant phenotype could be the parental trait, but it could also be the hybrid trait that exhibited only the parental trait and not the recessive one. They also emphasize that Mendel did not make any attempt to reduce the "meaning" of traits to underlying, cellular dispositions but considered it only with reference to traits and their appearance in succeeding generations.[25] Thus, to give an example, a yellow seed could be either the parental yellow trait itself or the hybrid yellow trait also including, but not exhibiting, the latent green trait. One could not tell from the yellow color itself what was the case, but one could do so from the traits of the subsequent generations (meaning that two hybrid "yellow" seeds would yield "green" offspring), as shown in Figure 1.2.

Mendel further explained this in the next section:

> Those forms which receive the recessive characteristic in the first generation do not vary any longer with respect to this characteristic in the second generation; they remain constant in their descendants. It is a different case with those that possess the dominating trait in the first generation. Of these [forms] two portions produce descendants which exhibit the dominating and recessive traits in the proportion 3:1 [and] thus show exactly the same behaviour as the hybrid forms; only one portion remains constant with the dominating trait.[26]

The first generation that Mendel mentioned here is actually what we usually refer to as the second filial generation or F_2: the offspring of the hybrids. Here he made a general observation: whereas those individuals that exhibit the recessive trait can only have offspring with that very same trait, among those who exhibit the dominant trait, there exists some differentiation. One-third of them will have offspring with the dominant trait, whereas two-thirds of them will have offspring with either the dominant or the recessive trait in a 3:1 ratio. When he analyzed the results, Mendel figured out that the ratio 3:1 resolved itself in all experiments into the ratio of 2:1:1, if the dominant traits were differentiated as parental or hybrid traits:

The proportion of 3:1 according to which the distribution of the dominating and recessive characteristic occurs in the first generation, thus resolves itself into the proportion 2:1:1 in all experiments, if one all along distinguishes the dominating trait in its meaning as a hybrid trait and as a parental characteristic. Since the members of the first generation originate directly from the seeds of the hybrids, *it now becomes evident that hybrids of two respective differing traits form seeds of which one half develops again the hybrid form while the other gives rise to plants which remain constant and receive the dominating and recessive character in equal parts.*[27]

It is easy to read too much here, in hindsight. Based on the typical Mendelian genetics teaching, when two heterozygotes, say *Rr*, are crossed, their descendants are expected to have the following phenotypic ratio: 3 round:1 wrinkled. However, the respective genotypic ratio is different: 1 *RR*:2 *Rr*:1 *rr* (see Figure 1.2). Many have inferred from the above quotation that Mendel thus "discovered" the segregation of alleles. But this is not what Mendel wrote about; he wrote that one-half of the offspring developed the hybrid form (combining both the dominant and the recessive trait), whereas the other half developed either the dominant or the recessive parental trait in equal proportions. This sentence indeed makes us think of what in Mendelian genetics is described as the "law of segregation" (Figure 1.2) but only because what we read fits well with what we already know. In what Mendel wrote there is no reference to any hereditary particles that are inherited independently and that unite after fertilization. He was actually even clearer about this right afterward:

If *A* denotes one of the two constant traits, for example the dominating one, *a* the recessive one, and *Aa* the hybrid form in which both are united, then the expression

A + 2Aa + a

yields the developmental series for the progeny of hybrids with two differing traits each.[28]

The expression *A + 2Aa + a* that Mendel used certainly reminds us of the one used in Mendelian genetics: *AA + 2Aa + aa*, which represents the segregation of alleles. But this was not what Mendel wrote about. The symbols he used represented traits and indicated how these traits would be passed on to the next generation. Thus, individuals "A" and "a" would have offspring that would remain—as Mendel put it—"constant" with respect to the trait. This means that they would exhibit the parental phenotype, whereas individuals "Aa" would have offspring in which the traits would segregate again (which is why theirs was described as a "hybrid" trait). Mendel observed the segregation of traits, not of any underlying factors.

Then Mendel looked into what happened when two different traits were considered. In one experiment, he crossed plants that differed in the shape and the color of the seeds. He thus transferred pollen from plants with wrinkled (angular in the translation by Müller-Wille and Hall) seeds and green albumen to a plant with round seeds and yellow albumen. The plants in the next generation (F_1) all had round seeds and yellow albumen. When these plants were crossed, 556 seeds were collected that had the following traits:

- 315 round and yellow
- 101 wrinkled and yellow
- 108 round and green
- 32 wrinkled and green[29]

Does this look familiar? Yes, it is similar to the 9:3:3:1 ratio of Figure 1.3. Here is what Mendel wrote about it:

> The developmental series thus consists of 9 members. Four of these appear once each in the series and are constant in both traits; the forms AB, ab resemble the parent-species, the other two represent the constant combinations between the conjoined traits A, a, B, b that are possible in addition. Four members appear twice each and are constant in one trait and hybrid in the other. One member occurs four times and is hybrid in both traits. The descendants of hybrids, once two kinds of differing traits are joined in them, therefore develop according to the expression:

$$AB + Ab + aB + ab + 2ABb + 2aBb + 2AaB + 2Aab + 4AaBb^{30}$$

Mendel thus described the traits of the offspring of hybrids not as the classic 9:3:3:1 ratio of Mendelian genetics (Figure 1.3) but as a 1:1:1:1:2:2:2:2:4 ratio, in which:

- two members resembled each of the parental plants (*AB* and *ab*)
- two members had one trait from each parental plant (*Ab* and *aB*)
- four members appeared twice and had a parental and a hybrid trait (*ABb, aBb, AaB, Aab*)
- one member appeared four times and had two hybrid traits (*AaBb*)

Once again, Mendel described his findings by reference to parental and hybrid traits. His conclusion was:

> It can therefore not be doubted that for all traits admitted to the experiments this statement possesses validity: *The descendants of hybrids, in which several essentially different traits are united, represent the members of a combination series, in which the developmental series for two differing traits respectively are conjoined.* At the same time this demonstrates, that *the behaviour of two respective differing traits in hybrid conjunction is independent of any other differences in the two parental plants.*[31]

This certainly brings to mind what in Mendelian genetics is called the "law of independent assortment". However, Mendel made reference to traits and not to any underlying factors.

To summarize, Mendel described the transmission of traits first in the hybrids (F_1) and then in their offspring (F_2). In particular, he observed that the hybrids obtained from the various crosses between different varieties were not always intermediate between the parental forms. Rather, some hybrids exhibited traits that were exactly the same as those of the parental plants. Mendel called the parental traits that appeared in the hybrids dominant and the parental traits that did not appear in the hybrids but reappeared in their offspring recessive. Thus, Mendel studied, and wrote about, the transmission of traits, not of hereditary particles. Neither did Mendel discover, as is commonly attributed to him, the segregation and the independent assortment of alleles. However, because of his experimental design and statistical analysis, he was able to observe the consequences of these phenomena.

Let us now consider in more detail why Mendel did not discover the laws of segregation and independent assortment of alleles, commonly attributed to him.

MENDEL DID NOT DISCOVER THE "LAWS" OF HEREDITY

In 1979, historian Robert Olby in a paper titled "Mendel No Mendelian?" clearly and convincingly showed that Mendel's paper was not about the laws of heredity as the stereotypical narrative of Mendelian genetics has it.[32] His aim was "to publicize a view of Mendel's work which strips it of inflated whiggish interpretations and places it squarely within the context of mid-nineteenth century biology".[33] Rather, it was about the study of hybridization. Olby showed this by providing different pieces of evidence from Mendel's own paper and from other writings. However, here I want to focus on what I consider to be the single most important and most convincing piece of evidence that clearly shows that "Mendel's fundamental conception was that of the "*independent and unchanged transmission of separate hereditary characters*" (i.e., traits).[34] Olby noted, as I have already pointed out, that "*It was not the factors or elements which had to be defined in the Versuche as dominating, latent, constant or varying, but the characters*".[35] Let us see why.

Recall from a previous section that the classic 9:3:3:1 ratio emerges from the combination of two 3:1 ratios. A cross between two heterozygotes ($Aa \times Aa$) will yield the following genotypic ratio in their offspring: $1AA:2Aa:1aa$, which corresponds to a phenotypic ratio of 3[A]:1[a] (where [A] and [a] stand for whatever phenotypes the alleles A and a correspond to). This would also be the case for the cross between two heterozygotes for another trait (say, $Bb \times Bb$), which would yield offspring with the genotypic ratio $1BB:2Bb:1bb$ and a phenotypic ratio of 3[B]:1[b]. So, when two individuals that are heterozygotes for both traits ($AaBb$) are crossed, what happens is that we can have the combination of the aforementioned results, as shown in Table 1.1

If we sum up what kinds of phenotypes we have in Table 1.1, what we get is 9[AB]: 3[Ab]:3[aB]:1[ab], or the classic ratio of 9:3:3:1. However, as we saw in the previous section, this ratio does not appear in Mendel's paper; rather the ratio he presented was a 1:1:1:1:2:2:2:2:4 one. Where does this come from? Olby argued that Mendel

---

TABLE 1.1

Genotypes (and Phenotypes within Brackets) of the Crosses between Two Individuals, Heterozygote for Two Traits

	AA	2Aa	aa
BB	$AABB$ → [AB]	$2AaBB$ → 2[AB]	$aaBB$ → [aB]
2Bb	$2AABb$ → 2[AB]	$4AaBb$ → 4[AB]	$2aaBb$ → 2[aB]
Bb	$AAbb$ → [Ab]	$2Aabb$ → 2[Ab]	$aabb$ → [ab]

TABLE 1.2
**Phenotypes of the Combination of the Crosses
between Individuals Differing in Two Traits**

	A	2Aa	a
B	AB	2AaB	aB
2Bb	2ABb	4AaBb	2aBb
B	Ab	2Aab	ab

wrote in his paper that it is the traits that are combined in various ways, already quoted in the previous section: *"The descendants of hybrids, in which several essentially different traits are united, represent the members of a combination series, in which the developmental series for two differing traits respectively are conjoined".*[36] Olby suggested that this can be illustrated with the representation in Table 1.2.

This is how the ratio $AB + Ab + aB + ab + 2ABb + 2aBb + 2AaB + 2Aab + 4AaBb$[37] emerges.

Therefore, by comparing Tables 1.1 and 1.2, it becomes clearly evident that Mendel did not describe the transmission of hereditary elements (such as genes) but rather that of phenotypic traits. Mendel distinguished between those traits that were common between the plants crossed and those that were not and suggested that the differentiating ones united in the hybrids whereas the common ones were transmitted unchanged. In other words, Mendel suggested a different pattern of transmission for these traits. Here is how Olby described Mendel's relation to Mendelian genetics:

> If we arbitrarily define a Mendelian as one who subscribes explicitly to the existence of a finite number of hereditary elements which in the simplest case is two per hereditary trait, only one of which may enter one germ cell, then Mendel was clearly no Mendelian. On the other hand, if by Mendelian we mean one who treats hereditary transmission in terms of independent character-pairs, and the statistical relations of hybrid progeny as approximations to the combinatorial series, then Mendel was a Mendelian.[38]

In my view, this excellently clarifies why Mendel's work did not have an impact before 1900 and why it did after 1900. Mendel did not provide a general theory of heredity in 1866; however, he developed an experimental approach that would later provide the foundation of genetics. How the latter happened is discussed in the subsequent chapters.

But was Mendel interested in heredity at all? Historian Garland Allen has argued that as hybridization is about mixing reproductive cells of different kinds and studying what the outcome of this is, and therefore how these cells interact, there has to be some underlying conceptualization of heredity. The distinction between hybridization and heredity is therefore confusing because the two processes are complementary.[39] But, I would argue, having some conceptualization of heredity is not the same as trying to develop a theory of heredity. Mendel might have speculated about this. But the most significant feature of his work was that he prioritized empirical work and evidence, and as we have seen his findings were about the traits that the hybrids and their descendants exhibited. Therefore, as Allen also pointed out, it is correct for historians to emphasize that Mendel was interested in establishing the laws of hybridization rather than developing a general theory of heredity. But, if this is so, what are those "factors" and "elements" that Mendel referred to in his paper?

> We therefore have to consider it as necessary that, also in the generation of constant forms on the hybrid plant, completely identical factors act together.[40]

> This development occurs according to a constant law, which is grounded in the material constitution and arrangement of the elements that attained a viable union in the cell.[41]

Isn't this what we see in Mendelian genetics? The constant forms are homozygotes (say, *AA* or *aa*), and they are generated by the action of identical factors acting together (a.k.a. alleles). The same is the case with elements. How elements (again alleles) are arranged within cells provides the grounds for development. Several scholars have suggested this. For instance, it has been argued that Mendel's interpretation was based on material hereditary elements that underwent segregation and independent assortment, citing the second quotation above.[42] Others have further suggested that Mendel's reference to "potentially formative elements" and to the "enforced association of this elements" demonstrated an understanding of the existence of pairs of hereditary elements.[43] Isn't this clear in the aforementioned quotations?

Well, not necessarily. According to Müller-Wille and Hall, Mendel adopted the terminology of Gärtner, who had used the term "Faktoren" to refer to the essence, or types, of the two species hybridized (see Chapter 2), rather than to the underlying dispositions for particular traits.[44] The term "Elemente" is also taken from Gärtner and seems to refer to the dispositions for certain parental traits. It could refer to particles, but it does not have to; it can refer only to a component of a whole. The important point is that rather than inferring from Mendel's statements what he might have meant, it is more informative to look at where he got his insights from. In his heavily annotated copy of Gärtner's book, Mendel had underlined phrases in which these terms are used, and so these are key in interpreting what Mendel had intended to state.[45] In their critical analysis of Mendel's paper (based on Sherwood's translation), botanists Alain Corcos and Floyd Monaghan suggested that Mendel never defined what these elements were, and this was a weakness in the explanatory scheme he developed. Mendel had written: "If one succeeds in joining a germ cell with a pollen cell of a different kind, then we must suppose that some compromise is taking place between those elements of the two cells that condition the mutual differences". But even if he correctly believed that there was something within cells the expression of which determined those traits, insofar as he did not have a clear idea of what that was, he could not explain the supposed "compromise" taking place between the elements of reproductive cells of different kinds.[46] It was the "mediation cell" thus emerging that would form "the foundation of the hybrid organism, whose development necessarily occurs according to a law other than that in the case of each of the two parent-species".[47]

Furthermore, Mendel also wrote about these elements:

> In the formation of these cells, all the elements participate in a totally free and uniform arrangement, while only the differing ones mutually exclude each other. In this way, the formation of as many kinds of germ and pollen cells would be enabled as there are combinations allowed for by the elements capable of development.[48]

What did Mendel mean by writing that only the differing elements mutually exclude each other? According to Müller-Wille and Hall, this means that "fertilization cells" could not end up containing elements of a different kind, in contrast to the "foundation cell" formed by their union. They also pointed out that having each fertilization cell receiving an equal amount of elements indicates that these were not particulate.[49] Corcos and Monaghan have interpreted the statement that only the differing elements mutually exclude each other, as meaning that Mendel assumed that the elements of the same kind did not separate. So, if Mendel thought of elements as alleles, then he would be stating that similar alleles (e.g., those found in homozygotes) do not segregate whereas only different ones (e.g., those found in heterozygotes) do. But this would be a self-contradiction for the law of segregation as we know it.[50] And it is probably not one, because Mendel wrote about the segregation of traits, not alleles.

It was actually William Bateson's interpretation of Mendel's paper that produced the notation that has been used ever since in Mendelian genetics. Whereas Mendel used single letters to represent trait combinations, it was Bateson who transformed this in a way that represented alleles. As Olby has shown, Mendel developed his notation without any concept of underlying factors or elements (alleles), and it was actually Bateson who was the first to point out its serious limitations. In commenting on Mendel's results with *Phaseolus multiflorus*, Bateson added a footnote about Mendel's way of symbolizing compound traits. Mendel had written:

> If the flower colour A were a combination of the individual characters $A_1 + A_2 + \ldots$ which produce the total impression of a purple colouration, then by fertilisation with the differentiating character, white colour, a, there would be produced the hybrid unions $A_1 a + A_2 a + \ldots$ and so would it be with the corresponding colouring of the seed-coats.[51]

Bateson added a footnote to this sentence in which he wrote:

> It appears to me clear that this expression is incorrectly given, and the argument regarding compound characters is consequently not legitimately developed. The original compound character should be represented as $A_1 A_2 A_3 \ldots$ which when fertilised by a_1 gives $A_1, A_2, A_3 \ldots a$ as the hybrid of the first generation. Mendel practically tells us these were all alike, and there is nothing to suggest that they were diverse. When on self-fertilisation, they break up, they will produce the gametes he specifies; but they may also produce $A_1 A_1$ and $A_2 A_2$, $A_1 A_2$ a &c., thereby introducing terms of a nature different from any indicated by him. That this point is one of the highest significance, both practical and theoretical, is evident at once.[52]

What does all this entail? First of all, that the Mendelian genetics we teach today, with the 3:1 and the 9:3:3:1 ratio, is actually Batesonian genetics. We do not teach Mendel's understandings but rather the reinterpretation of his findings by Bateson. As we see in Chapter 6, by defending Mendel's work against criticisms, and by making the resulting model the prevailing one, it was Bateson who established what came to be known as Mendelian genetics as a discipline. In this sense, all the negative consequences that I present in Chapters 8–10 as being due to the naive Mendelian genetics models are actually models established by Bateson that should be better described as Batesonian. Mendel could not have known what geneticists in the early 20th century came to know. Interpreting what we think he meant in hindsight, and thus attributing to him an understanding of heredity based on the existence of hereditary particles, can rely only on a strongly anachronistic reading of his paper. This has caused much confusion. The stereotypical representation of Mendel as a heroic, lonely pioneer who discovered the laws of segregation and independent assortment of alleles distorts not only the actual history but also how science in general is actually done.

Strictly speaking, Mendel did not discover the laws of segregation and independent assortment commonly attributed to him; nonetheless, because of his experimental design, he was able to observe their consequences under particular experimental conditions. Neither did he discover that inheritance is particulate in nature, and this is why he did not write about any hereditary particles in his paper. Rather, Mendel investigated how particular characters developed in hybrids and their progeny. Mendel's main conclusion was that all sperm and eggs contained a single version of each of their characters. When a sperm fused with an ovum containing a different version of a particular character, the resulting offspring would eventually contain both versions, which were independently transmitted to the sperm and eggs that the offspring would later produce. In short, Mendel studied the transmission of characters in hybrids and their progeny, not of any factors or genes.

Mendel was trying to provide answers to questions related to plant breeding and hybridization, to which we now turn.

NOTES

1 Biggs et al. (2009), p. 277.
2 Miller and Levine (2010), p. 262.
3 Nowicki, (2012), p. 167.
4 Reece et al. (2012), p. 262.
5 Ward et al. (2008), p. 98.
6 Walpole et al. (2011), p. 86.
7 Griffiths et al. (1999), p. 104.
8 Audesirk et al. (2002), pp. 212, 218.
9 Raven et al. (2008), pp. 220, 222.
10 Mader (2004), p. 182.
11 East (1923), p. 234.
12 Orel (1984) chapters 1,2; Orel (1996) Ch. 2. In the present book, you will mostly find information that is directly related to the results of Mendel's work that he presented in 1865. For biographical information, there are several different biographies of Mendel available.

The biography that most expert-historians consider as the standard was written by Vitězlav Orel and was published in 1996 (there is another, concise one by Orel, published in 1984). Orel played a key role in the organization of the Mendel symposium of 1965 in Brno and in promoting Mendel's work more broadly. With Jaroslav Kříženecký, he cofounded the Mendelianum in Brno, which served as an intellectual bridge between the East and West during the Cold War. He also served for almost 30 years as the editor of the journal *Folia Mendeliana* (Paleček, 2016). Besides Orel's, however, there a few other biographies that I occasionally use throughout the present book. One is the first-ever biography of Mendel written by Hugo Iltis, who interviewed people who had actually met Mendel. This was published in German in 1924, but I am drawing upon the English translation published in 1932. Another is the first volume of a rich and comprehensive account of Mendel's life by Jan Klein and Norman Klein (published in 2013; the second volume is mentioned therein but has not been published yet). A concise account of Mendel's life and work in also given by Daniel Fairbanks in a book published in 2022. Finally, there is a 2022 biography (perhaps better iconography), by Anna Matalova, Emeritus Head of the Mendelianum of the Moravian Museum in Brno, and her daughter Eva, which provides previously unknown documents and other materials related to Mendel's life and work.

13 Mendel (2020), p. 29.

14 Orel (1984), pp. 42–45; Orel (1996) Chapter 5.

15 There are several, old and recent, translations of Mendel's paper in English. (For a detailed account, see the introduction by Müller-Wille and Hall in Mendel (2020), pp. 14–16.) The first was the one by Charles Druery for the journal of the Horticultural Society of London, a modified version of which was made widely available by William Bateson in his 1902 book about Mendel. Another one is the translation by Eva Sherwood, published along with several other key documents, in 1966 (Stern and Sherwood, 1966). A more recent translation was published in 2016 by Scott Abbott and Daniel Fairbanks in the journal *Genetics* (Abbott and Fairbanks, 2016), also included as an Appendix in Fairbanks (2022). However, in the present book, I rely on the translation made by historians Staffan Müller-Wille and Kersten Hall. The reason for choosing this translation is that, as they explain in their introduction, their translation is accompanied by detailed notes that consider alternative readings, justify the choices of words made, and thus provide a guide to the original German text for those who cannot read German. In this way, this translation allows readers to decide for themselves how to interpret what Mendel wrote (Mendel, 2020, pp. 16–18). Most of the clarifications that I provide throughout the book about the use of terms in the quotations from Mendel's paper are those by Müller-Wille and Hall in their comments.

16 Mendel (2020), p. 23.

17 Mendel (2020), p. 29.

18 Mendel (2020), p. 53.

19 Mendel (2020), p. 75.

20 Müller-Wille and Hall explain that "dominating" is a more accurate term than "dominant", because it denotes an activity and not a state. Nevertheless, as the term broadly used in Mendelian genetics is "dominant", I am also using that in the present book.

21 Mendel (2020) p. 35.

22 Mendel (2020) p. 39.

23 Mendel (2020) p. 31.

24 Mendel (2020), p. 39.

25 Mendel (2020), p. 43.

26 Mendel (2020), p. 45.

27 Mendel (2020), p. 47, emphasis in the original.

28 Mendel (2020), p. 49.

29 Mendel (2020), p. 55.

30 Mendel (2020), p. 57.
31 Mendel (2020), p. 59, emphasis in the original.
32 Olby (1979); Olby's book *Origins of Mendelism* was published in 1966. In 1985, a second considerably expanded edition was published, which included his 1979 paper. All quotations from Olby (1979) in the present book refer to the respective pages in Olby (1985).
33 Olby (1985), p. 235.
34 Olby (1985), p. 247, emphasis in the original.
35 Olby (1985), p. 248, emphasis in the original.
36 Mendel (2020), p. 59, emphasis in the original.
37 Mendel (2020), p. 57.
38 Olby (1985), p. 254
39 Allen (2003), pp. 65–66.
40 Mendel (2020), p. 63.
41 Mendel (2020), p. 89.
42 Hartl and Orel (1992), p. 249.
43 Fairbanks and Rytting (2001), p. 745.
44 Olby (1985), pp. 32–33; Mendel (2020), p. 121.
45 Mendel (2020), p. 138.
46 Corcos and Monaghan (1993), p. 164.
47 Mendel (2020), p. 89.
48 Mendel (2020), p. 89.
49 Mendel (2020), p. 140.
50 Corcos and Monaghan (1993), p. 174.
51 Bateson (1902), p. 80.
52 Bateson (1902), p. 80.

2 Mendel Was a Brilliant Experimentalist of His Time

MENDEL WAS NOT AHEAD OF HIS TIME

The stereotypical narrative of Mendelian genetics presents Mendel as being ahead of his time. But he was neither the first to work on peas nor the only one among his contemporaries to study hybridization. The first person to select peas for a systematic study of the transmission of traits was Thomas Andrew Knight, an insightful horticulturalist in Britain who worked with hybrids. In 1823 he presented a paper to the Horticultural Society of London, of which he was president between 1811 and until his death in 1838. The title of the paper was "Some Remarks on the Supposed Influences of the Pollen, in Cross-Breeding, on the Color of the Seed-Coats of Plants, and the Qualities of their Fruits". Here is what Knight wrote:

> The numerous varieties of strictly permanent habits of the Pea, its annual life, and the distinct character in form, size, and colour of many of its varieties, induced me, many years ago, to select it for the purpose of ascertaining, by a long course of experiments, the effects of introducing the pollen of one variety into the prepared blossoms of another. . . . When, in my experiments, the pollen of a gray Pea was introduced into the prepared blossoms of a white variety, no change whatever took place in the form, or colour, or size, of the seeds; all were white, and externally quite similar to others which had been produced by the unmutilated blossoms of the same plant. But these when sown in the following year, uniformly afforded plants with coloured leaves and stems, and purple flowers; and these produced gray peas only. When the stamens of the plants which sprang from such gray peas were extracted, and the pollen of a white variety, of permanent habits, was introduced, the seeds produced were uniformly gray; but many of these afforded plants with perfectly green leaves and stems, and with white flowers, succeeded, of course, by white seed.[1]

Looks familiar? Knight crossed plants with gray seeds with plants with white seeds. In the plants of the first generation, all seeds were gray. However, in the plants of the second generation, there were both plants with gray seeds and plants with white seeds. What we see here is that Knight, long before Mendel, was aware of the properties of peas for experimentation. Furthermore, he arrived at results similar to those of Mendel, even though he did not count the seeds or report the ratios as meticulously as Mendel did. Knight had actually observed dominance as early as 1799.[2] Now you may wonder: was Mendel aware of Knight's work? As he is not mentioned in Mendel's paper, and as we know that Mendel could not read English, it could be the case that he was not. However, Gärtner, whose work Mendel studied carefully, did

DOI: 10.1201/9781032449067-3

cite the work of Knight so it is possible that Mendel knew of his work, even if he never studied it in detail.[3]

It is also important to note that Mendel was not alone among his contemporaries in doing the kind of work he did. During the 1850s, Charles Naudin at the National History Museum in Paris undertook extensive experiments in hybridization, being interested in its evolutionary significance. One topic of interest was the distinction between varieties and species. He corresponded with Darwin between 1862 and 1882, and they exchanged publications. Between 1852 and 1861, Naudin carried out experiments that aimed to show that reversion, the reappearance of ancestral forms, was possible. When in 1860 the Academy of Sciences proposed plant hybridization as the topic for a prize, Naudin submitted an essay based on his numerous experiments done by that time and won.[4] The essay, written in 1862 and published in a complete form in 1864, was titled "New Research in Hybridity in Plants ("Nouvelles recherches sur l'hybridité aux végetaux"). There Naudin wrote:

> All hybridists agree that the hybrids (and these are always first-generation hybrids) have mixed forms, intermediate between those of the parental species. This is indeed what happens in the vast majority of cases, but this does not entail that these intermediate forms are always at an equal distance from those of the two species. We have often noticed, to the contrary, that they are sometimes much closer to one than the other. . . . We have also noticed that the hybrids sometimes resemble one of the two species more than the other in certain parts.[5]

Naudin continued:

> In my view, these differences in resemblance, sometimes very large, between the hybrid and its parents are mostly due to the significant preponderance that many species show in their crosses. . . . From the second generation on, the appearance of hybrids changes in the most remarkable way. Normally, the perfect uniformity of the first generation is succeeded by an extreme variation of forms, some approaching the specific type of the male parent, the others that of the female parent, whereas others returning suddenly and completely to one or the other form. . . . It is indeed in the second generation that, in the majority of cases (and perhaps in all of them), the dissolution of the hybrids forms begins.[6]

Sounds familiar? Naudin clearly observed the uniformity of hybrids in the first generation and the segregation of characters in the second. But it seems that he never calculated frequencies or did the kind of mathematical analysis that Mendel did. Here is how he explained why this segregation happened:

> All the hybrids that I have observed up to the second generation have shown me these changes and have manifested the tendency to revert to the forms of the species from which they were produced. . . . All these facts are naturally explained by the *disjunction of their two specific essences* in *the pollen and the ovum of the hybrid.*[7]

Naudin wrote explicitly about the causes of segregation, suggesting that the hybrid was a live mosaic, consisting of invisible elements; those of the same species had higher affinities for one another and thus tended to get together, forming considerable masses even whole organs.[8]

We see therefore that among his contemporaries, Mendel was not the only one to make the observations about the uniformity of characters in the hybrids and their segregation in the next generation. Naudin even speculated about the underlying causes of the forms he observed, the "elements" responsible for the observed characters. This of course does not make Naudin the "father of genetics" any more than Mendel is. It just shows that the questions that preoccupied Mendel were also the preoccupations of other naturalists of his time. In all cases, we should refrain from interpreting such speculations in hindsight.

Mendel was a scientist of his time, and there were particular conditions and people who had influenced him. Among Mendel's influences, it is worth highlighting three: the socioeconomic context and the breeding practices in Brno and Moravia more broadly; the hybridist research tradition of the late 18th and early 19th centuries; and the academic botany in Vienna.

THE INFLUENCE OF THE MORAVIAN SOCIOECONOMIC CONTEXT AND BREEDING PRACTICES

The stereotypical narratives about the great discoveries made by the great geniuses in science tend to focus on the geniuses themselves and overlook their cultural and social milieu. A careful consideration of history more broadly makes this clear. A typical example is Charles Darwin and his theory of evolution by natural selection. He was a brilliant and hard-working naturalist for sure. But he also had available many intellectual and practical resources, characteristic of the Victorian era, which were crucial for the development of his theory: Thomas Malthus and Adam Smith were political economists who developed their theories in the English context; animal and plant breeding was a form of Victorian technology; John Herschel and William Whewell were among the first philosophers of science in a tradition based on Newton's science; and the Anglican Church had a tradition, natural theology, that put emphasis on the idea of adaptation. All these were important intellectual resources. But there were also important practical resources due to the industrialization and imperialism of the British Empire, which made Darwin's research possible. He famously travelled around the world for five years aboard the HMS Beagle. Thanks to the British ships traveling around the world, he also developed a vast network of correspondents. It thus may be no coincidence that Alfred Russel Wallace, who lived at the same time and in the same culture as Darwin, independently came up with a similar theory.[9] The socioeconomic context of Brno was similarly crucial for Mendel's work.

Since the beginning of the 19th century, Brno had become a pioneering center of animal and plant breeding practices. A key figure for these developments was Christian Carl André, who moved to Brno in 1798 to work as a teacher. His interests at the time were many, but mineralogy was an important one, and it was another reason for his move to Moravia. There he made acquaintances with other people with similar interests, one of whom was the Count Hugo Franz Algraf zu Salm-Reifferscheidt (hereafter simply referred to as Salm), who was an aristocrat but also an entrepreneur in the textile industry. In 1806, they reorganized the Moravian Agricultural Society, which they merged with the private Natural Science Society, thus producing the Agricultural Society, with Salm as its president and André as its secretary. One of the main objectives of this society was the improvement of wool production, which was an important financial resource for the area. In parallel, André developed a journalistic activity that focused on sheep breeding, with the aim of producing better and more wool. In 1811, Salm appointed André as his economic advisor, which allowed the latter to give up teaching.[10]

Following the work of Robert Bakewell in England, Baron Ferdinand Geisslern successfully developed sheep breeding (but his interest was in the production of fine wool, whereas the British were mostly interested in the production of meat). Following Bakewell's methods, which included inbreeding such as mating mother with son and father with daughter, Geisslern tried to perfect sheep breeding. His success gained him the title of the "Moravian Bakewell". However, whereas inbreeding was a typical animal breeding practice in England, it was met with hesitation in continental Europe on religious grounds. Baron J.M. Ehrenfels, a prominent Austrian breeder, and others, had argued that producing animals by inbreeding led to the deterioration of the stock as a whole, and so thought André. Ehrenfels was concerned about the deterioration of the quality of wool, which he attributed to three factors: (i) selection of animals to breed on the basis of their physical appearance; (ii) an emphasis on quantity rather than quality; (iii) an overemphasis on inbreeding. These debates led to various discussions about breeding and how to improve it. André's son, Rudolph, who had published a textbook on the improvement of sheep breeding in 1816, was also instrumental in the formation of the Sheep Breeders Society. Eventually, comparisons among different kinds of wool in a meeting of this society supported the conclusion that inbreeding indeed worked well.[11]

An interesting figure in these debates was the Hungarian Imre Festetics, who disagreed with Ehrenfels. André asked him to summarize the arguments in favor of inbreeding in an 1819 paper, in which he proposed what he described as the "Genetic Laws of Nature" ("Die genetischen Gesetze der Natur"). These were the following: (1) The traits of healthy animals tended to be inherited by their offspring; (2) animals tended to exhibit traits not exhibited by their parents but by their grandparents or more remote ancestors; (3) the animals that possessed the desired traits might have offspring with other traits that were undesired by the breeders; and (4) inbreeding was successful in producing offspring with the desired traits only when the animals mated were carefully chosen so that they exhibited the desired traits in abundance.[12] It might be tempting to identify another forefather of genetics here and one who used the actual term already in 1819. But it is sensible to refrain from doing so. Having said that, the fact that this kind of theorizing and discussion was taking place in Brno was very important for Mendel's work.

Equally important was the fact that Napp, the Abbott of Mendel's monastery, was involved in several breeding associations. In 1827, he became president of the Pomological and Oenological Society of Brno, and by the time the young Johann Mendel joined the monastery, in 1843, Napp was already quite influential. One of the outcomes of this influence was the appointment, in the same year, of Franz Diebl to give courses on agriculture at the Philosophical Institute in Brno, with Johann Karl Nestler giving similar courses at Olomouc University since 1823. Whereas Nestler preferred animal breeding, Diebl was interested in plant breeding. He emphasized that the careful selection of parents was crucial for the transmission of traits and suggested that peas and beans were ideal subjects for experimentation. After 1841, Diebl collaborated closely with Napp, with Mendel himself also attending his lectures. Both Nestler and Diebl taught about breeding, arguing that inbreeding was a useful means for getting the desired traits in the offspring.[13]

Nestler also considered heredity as a subject of research. Whereas in the past he had referred to "generation with heredity", in an 1836 meeting he changed his focus. Referring to wool production, he focused on the patterns of transmission of traits from parents to offspring, leaving aside the puzzling questions related to development and physiology. Not all people at the meeting were in agreement with Nestler's approach, but he was convinced that the Moravian breeders had detailed records for over 40 years that would provide insights into questions about heredity. He therefore asked members of the society to look at their records, from which he expected to acquire useful information in order to explain the patterns of heredity and thus create a new breeding system. Napp was a keen supporter of Nestler's ideas. However, they were both well aware that practical breeders might think of heredity as irrelevant to their goals; they had to be convinced that knowledge of heredity would provide them with better wool. In the 1837 meeting Napp asked the important question: "What we have been dealing with is not the theory and process of breeding. But the question should be: what is inherited and how?". For Napp the question was not if traits were inherited—this was already evident to experienced breeders—but what was behind an inherited trait.[14]

In an exemplar case of the kind of anachronism that ought to be avoided in the study of the history of science, Peter van Dijk, Franz Weissing, and T.H. Noel Ellis argued that

> Mendel's broad interest in plant biology was clearly sanctioned by Napp's comments relating to the need for a scientific study of inheritance. Although the word inheritance was used only once in the text of the Pisum paper and was missing from the title, the paper is unmistakably about the rules of inheritance.[15]

Anyone who has actually read Mendel's paper would be left wondering about this statement. How exactly is the paper clearly about heredity when the term, as we saw earlier, is not mentioned therein? As historians Müller-Wille and Hall have explained, Mendel never used the term "*Vererbung*", *the* German counterpart for heredity or inheritance. He did use the verb "*vererben*" a couple of times, but he did this in order to emphasize that a certain trait was *not* inherited. However, there are references to the transmission of traits, as Mendel used occasionally a German verb construction "*übergehen . . . auf*", for which there is no straightforward equivalent in English. In German, "*übergehen*" is often used in the context of inheritance of land property or titles, and this, according to Müller-Wille and Hall, may provide further evidence that Mendel's understanding of heredity was not the same as that of later geneticists. They also pointed out that whereas "*übergehen . . . auf*" has been previously translated as "are transmitted to", this may not be accurate as, being an intransitive verb, it neither has a passive voice nor can have a direct object.[16] Furthermore, if Mendel was influenced by Napp to set up a scientific study of heredity, why there was no reference at all in his paper to Napp or Nestler who had clearly argued for the importance of such a study? Van Dijk and colleagues do not address this question. Rather, they have reached the conclusion that Mendel was studying heredity and looked for evidence to support it, no matter what this is. They went as far as to quote a local newspaper article, stating: "We honor every endeavor to approach truth in a practical manner". From this statement only, and even though the article was negative about Mendel's horticultural work, Van Dijk and colleagues concluded: "This remark is intriguing, since it suggests that Mendel was empirically solving a scientific question, which in this instance would have been the understanding of the inheritance of traits".[17]

Such anachronisms notwithstanding, it remains a historical fact that Mendel joined the monastery in Brno at a time when animal and plant breeding was of central interest in the social and economic context in which he would live. This provided him with important intellectual and practical resources for his own work.

THE INFLUENCE OF THE RESEARCH TRADITION OF LATE 18TH- AND EARLY 19TH-CENTURY PLANT HYBRIDISTS

As we saw in Chapter 1, Mendel was clear in his paper that he was interested in studying hybrids. Historian Staffan Müller-Wille and Mendel's biographer Vítězlav Orel have convincingly shown that Mendel's work was in the same research tradition as, and also guided by, those of late 18th- and early 19th-century hybridists such as Carl Linnaeus, Joseph Gottlieb Kölreuter, and Karl Friedrich Gärtner.[18] It was Linnaeus who started this research tradition in 1751 with the first systematic study of plant hybrids, by addressing their formation as a regular, natural phenomenon that could be studied scientifically—without however conducting any experimental work himself. It was through this work that criteria for distinguishing between species and varieties were established. Since Linnaeus, species had been considered to consist of individuals that were related by descent and differed from other species in traits that remained constant, independently of external factors such as climate or soil. In contrast, varieties were considered to result from the effect of external factors on individuals of the same species, which in turn resulted in variable traits. Therefore, plants of the same species that had developed under exactly the same external conditions should be perfectly identical. If individual plants that had developed under stable, homogenous conditions exhibited a difference in one trait, while being perfectly identical in all others, they should be considered as representing a different species.[19]

According to Müller-Wille and Orel, it is precisely this Linnean definition of species that Mendel relied upon. As he wrote in his paper:

> If one wanted to apply the strictest definition of the species concept, according to which only those individuals belong to one species which under entirely identical circumstances also exhibit entirely identical traits, then no two of them can be counted as belonging to one species.[20]

According to historian Robert Olby, this definition entails that if any plant exhibited characteristics intermediate between those of two known species, it must have been a hybrid. Furthermore, Linnaeus assumed that this would have been a true-breeding hybrid and therefore a new species.[21] Mendel relied on this definition throughout his paper, and as Müller-Wille and Orel pointed out he repeatedly referred to his pea varieties as different "species" (Arten) in this sense, something that several 20th-century English translations have masked by translating "Art" as "variety", "kind", or "strain". Mendel clearly indicated that the 22 pea varieties he had selected for his experiments had been found to yield "identical and constant descendants throughout; at least in the two test years no essential alteration was to be noticed".[22] Thus, the Linnaean species concept provided the means of distinguishing among species on the basis of individual trait differences.[23] However, the view that hybridization could produce new species was unacceptable to other naturalists. If humans could produce new species indefinitely simply through the hybridization of existing species, this would distort the presumed harmony infused in nature during its creation.

In contrast to Linnaeus who performed no experiments, Joseph Kölreuter undertook extensive hybridization experiments—he was indeed the first one to do so—between 1760 and 1766. After receiving his degree from the University of Tübingen in 1755, he went to St. Petersburg, where he worked as a natural historian in the Academy of Sciences until 1761. It was there that Kölreuter likely became inspired to conduct his subsequent work, as in 1760 Linnaeus was awarded a prize by the Academy of Sciences of St. Petersburg for an essay he wrote on whether sexes existed in plants or not. Linnaeus argued that some parts of the plant were derived from the paternal plant, whereas others were derived from the maternal one, citing cases of plant hybrids to support this conclusion.[24] Kölreuter carried out more than 500 different hybridizations with 138 species and examined the shape, color, and size of the pollen grains from over 1,000 different species. He followed two lines of argument in order to demonstrate that the creation of new species from hybridization was impossible. On the one hand, Kölreuter concluded that F_1 hybrids for any given cross were similar to one another and most of their characteristics were intermediate between those of the parental species, and they did not exhibit a combination of parental traits as Linnaeus believed. Furthermore, their offspring (F_2) tended to resemble the parental species and not the F_1 hybrids; in other words, the offspring of the hybrids in some way reverted to the characteristics of the original, parental species. On the other hand, Kölreuter tried to show that true species hybrids were always sterile and thus could not reproduce, a possibility to which Linnaeus seems to have been indifferent. Kölreuter argued that hybrids between species did not produce fertile offspring, whereas although hybrid varieties were fertile, their traits were always intermediate between the parental traits and thus remained within the species' limits. Thus, for Kölreuter, this was evidence that hybrids did not breed true and that therefore hybridization could not produce new species as Linnaeus had believed. This also stood, in a sense, as a confirmation of the stability of species.[25]

When Kölreuter returned to Germany from St. Petersburg in 1761, he was able to continue his experiments in the private garden of botanist Joseph Gärtner, whose son Karl was inspired to also conduct experiments, focusing on fertilization and hybridizations in plants. One reason for this was the debates during the 1820s on the sexuality of plants, with August Henschel questioning the existence of sex in plants and the relevant work of Kölreuter. Another reason was the question of whether there was a direct color action of foreign pollen on the coats of the resulting seeds, with some previous experiments by Andrew Thomas Knight and Alexander Seton demonstrating such an effect, whereas others by John Goss did not. Thus, Gärtner began his experiments in 1824 crossing varieties of maize with different seed colors; in one case he obtained in the F_2 a ratio of 3.18:1 between grayish-red and reddish-gray seeds, which of course brings to mind Mendel's work. But neither Gärtner nor anyone else thought in terms of segregation of traits. In 1825, Gärtner repeated the experiments of Kölreuter with tobacco plants, and the results strengthened his conviction, in agreement with Kölreuter, that it was the total nature of the species, not that of the separate organs, which determined the direction and form that the hybrids took. In fact, they both considered the marked infertility and tendency of tobacco hybrids to revert to the original types in good agreement with their view that species were stable and distinct. In 1829, Gärtner crossed pea varieties with differently colored seeds, obtaining both kinds of results. In 1849 he published the results of about 10,000 artificial fertilizations involving 700 species, and yielding 250 hybrids, in a book in German, the title of which would be *Experiments and Observations upon Hybridization in the Plant Kingdom* (the title in German was *Versuche und Beobachtungen über die Bastarderzeugung im Pflanzenreich*). If you are wondering how Gärtner managed to conduct such a huge number of experiments with a large number of species, it is worth noting that, in contrast to Kölreuter, he was independently wealthy, thanks to his family and that of his wife, and so he was able to acquire several gardens spanning a large area in his hometown—he also paid the expenses for the printing of his book himself.[26] Darwin had acknowledged the importance of Gärtner's book, which was a very important source for his own writings on hybridization. In a 1864 letter to his botanist friend Joseph Dalton Hooker, he remarked: "I wish the Ray Soc[iety]. would translate Gärtner's Bastarderzeugung; it contains more valuable matter than all other works put together, & would do great service if better known".[27]

A key difference between Gärtner and Kölreuter, as well as Linnaeus, is that Gärtner did not believe that there were any particular criteria by which species can be distinguished from varieties. However, they all accepted an essentialist view of species, initially established by Linnaeus. The species essence was expressed in the hereditary traits of a plant; however, it was not a mere composite of those traits but a unit. Gärtner described that unit as a "factor". During hybridization, the two factors of the hybridized species were brought together and were expressed in the hybrids to a greater or lesser extent. Gärtner believed that how many offspring of each kind would be produced depended on the sexual relationship between the two species, which he quantified as follows: the number of seeds from the crossing of species A × B/the number of seeds from the self-crossing of species A. He called this fraction "elective affinity", and he considered it as an expression of the essence of a species.[28] On this basis, Gärtner proposed a "natural classification" of hybrids, which was exclusively based on the composition and descent of hybrids. Instead of relying on empirical generalizations, as Linnaeus and Kölreuter had done, Gärtner's classification of hybrids reflected the combinatory operations used in his hybridization experiments. This classification scheme could therefore serve as an instruction for systematically conducting hybridization experiments. It should be noted that Gärtner's conception of pure species was one that entailed that they would remain constant insofar as external conditions remained the same. When they were brought together with other pure species in hybridization, again insofar as external conditions remained the same, the outcome would always be the same hybrid forms, which would in turn indicate their inner nature.[29]

Why does all this matter? We know that Mendel read in detail Gärtner's book, as his heavily annotated copy is found in the Mendel Museum in Brno.[30] Philosopher Yafeng Shan has argued that Gärtner's influence on Mendel is clearly evident in their common terminology, as they both used the same terms in the same sense. The terms *Entwicklung* (development) and *Hybriden* (hybrids) are central terms in both Gärtner's and Mendel's work, thus indicating a common interest in hybrids and their development.[31] Three of Gärtner's insights mentioned earlier are crucial for understanding Mendel's work, all three of them being reflected in Mendel's paper:

1. The classification of plant hybrids, based on pure species from which hybrids could emerge;
2. The uniformity of simple hybrids, which resulted from the hybridization of pure species; and
3. The idea that there is an inner unit, the factor, that reflected the species' essence.[32]

THE INFLUENCE OF VIENNA'S ACADEMIC BOTANY

Mendel enrolled as a student at the University of Vienna in 1851. There he attended the lectures of Christian Doppler and later of Andreas Ritter von Ettingshausen, who succeeded Doppler after his untimely death in 1853. It seems likely that Mendel was influenced by both of them: Doppler's mathematics textbooks published during Mendel's studies explained the use of combinatorial theory and probability theory in science, whereas Ettingshausen emphasized the importance of the mathematical generalizations of experimental results. But perhaps the greatest influence on Mendel came from the newly established courses of botany. Morphology and plant systematics were taught by Eduard Fenzl, who was the director of the botanical gardens (and also the grandfather of Mendel's "rediscoverer" Erich von Tschermak-Seysenegg—see Chapter 6), whereas the new subjects of experimental plant anatomy and plant physiology were taught by Franz Unger (Figure 2.1). It is probably from Unger that Mendel came to know about the debates over plant hybridization, already discussed in this chapter.[33]

Unger did not restrict himself to university teaching but also tried to educate the general public. In 1852, he began publishing an anonymous series of *Botanical Letters*, 17 overall, in the local newspaper *Wiener Zeitung*.[34] These letters were subsequently published as a book in 1852, an English translation of which appeared in 1853.[35] Unger argued that species were not fixed and that the world of plants had emerged gradually. Mendel was his student at that time, so he probably became familiar with Unger's views according to which variants arose in natural populations that could give rise to varieties and subspecies and sometimes species. This view stood in contrast to the views of Kölreuter and Gärtner, who, as we already saw,

a)   b)

FIGURE 2.1 (a) Franz Unger (1800–1870), in 1853. (Courtesy: Archive of the University of Vienna, picture archive. Originator: Adolf Dauthage, Druck: J. Höfelich Signatur: 135.735); (b) Eduard Fenzl (1808–1879) in 1853 (Courtesy: Archive of the University of Vienna, picture archive. Originator: Adolf Dauthage, Druck: J. Höfelich Signatur: 135.628).

had both come to the conclusion that hybridization could not produce new species.[36] Perhaps it was this question that aroused Mendel's interest and made him perform his hybridization experiments when he returned to Brno in 1853. However, as we saw, understanding hybridization was any way of interest to the breeding practices in the socioeconomic context of Brno.

The anonymous set of the *Botanical Letters* appeared between May 28 and October 18, 1851 (but here I am drawing on the 1853 English edition). In letter XVI titled "Chronological Aspect of the Plant—History of the Plant-World", Unger made some evolutionary claims, bold for the time and place, arguing that the world of plants "has gradually developed itself step by step".[37] This gradual evolution (he did not use this term though) was not linear but branching:

> Thus the foundation of the whole plant-world is not by any means a one-sided lineal progression, but a radiation broadening out on all sides, and in their primitive shapes is contained the whole of the existing vegetation, as though in a chalk-drawing made from meagre outlines.[38]

Unger thus explicitly questioned the stability of species:

> Thus proceeds the ideal of the plant, as at first from cell to cell, from leaf to leaf, from sprout to sprout, from individual to individual, here also in continuous destruction and

new-birth of genera in unbroken undulation of renewal, one epoch of creation determining the other, each new, each strange, each proceeding from the earlier elements
modified, but even ennobled.[39]

Not surprisingly, these evolutionary aspirations did not go unnoticed. Sebastian
Brunner, the editor of the Catholic newspaper *Wiener Kirchenzeitung*, assumed that
the author of these anonymous letters was Unger and decided to react. In an editorial
in his newspaper, titled "Our Universities" and published a week after the publication
of the last of the *Botanical Letters*, Brunner expressed his concerns for the "pagan-
ist" views taught in universities. He was primarily concerned with the moral, not the
physical, distinction between humans and other animals. But his criticism of Unger
was relatively mild, and Unger generally refrained from replying to such criticisms.
However, when a few months later the *Botanical Letters* were published together as
a book, under Unger's name, Brunner criticized him again. In an editorial published
on January 29, 1852, Brunner wrote that Unger would rarely go to church and that he
held a pantheistic worldview. On January 4, 1856, Brunner described Unger as the
Austrian representative of scientific or metaphysical materialism, which claimed that
it was possible to explain the mind in terms of matter and material forces. Such views
and attitudes on behalf of Unger were, according to Brunner, at odds with being a
professor at the University of Vienna that had the goal to disseminate the Catholic
belief. Because of this attack, it seemed that Unger would have to resign. He filed
a lawsuit against Brunner's newspaper but did not respond to any of the criticisms
in print. However, a petition signed by 401 students in the medical faculty that was
sent to the minister of education, Graf Leo Thun, asserted that Unger had never
addressed religious issues in his classes. Unger eventually published a clarification
of his views, denying any support for pantheism or materialism and asserting that
his scientific work never contradicted the Christian belief in one personal God. This
made Brunner drop the issue.[40]

Unger's appointment at the University of Vienna took place right after the 1848
revolution in the Austrian Empire that had brought about significant social changes.
One of these was the adoption of a university system that prioritized scholarship for
the sake of knowledge itself rather than for practical applications, as well as academic
freedom that allowed professors autonomy in their research, teaching, and publishing. This is what made writing and publishing the *Botanical Letters* possible in the
first place. Before the reforms, for instance, when Unger himself was a student at the
University of Vienna back in the early 1820s, the Imperial Educational Commission
determined what each professor would teach and which books they would use, and
its permission was required for professors to use their own materials. But the new
context after 1848, with the Educational Commission having been replaced by a
Ministry of Education, allowed Unger to publicly express his evolutionary views. His
evolutionary theory was developmental in nature, in the sense that it relied on finding
analogies between embryonic stages and ancestral forms, assuming that the same
forces formed both the embryos and the species. He also argued that present-day bio-
geographical distributions were best explained historically, in an evolutionary way.
Whereas initially Unger accepted the idea of spontaneous generation, and the gradual replacement of older species by new ones, the developments in cell theory of the

1840s made him realize that cells always arose from preexisting ones and therefore switch to a view of universal common descent, as is evident in the quotations from his *Botanical Letters* above. The idea of material continuity between generations of cells, even when parent and offspring differed radically in form, became the basis for Unger's revised theory of 1851 and 1852, which held that all plant species were related by common descent.[41]

Historian Robert Olby has argued that similarities in the experimental programs of Unger and Mendel might suggest that Mendel's interest in the causes of variability was due to the teaching of Unger.[42] Historian Sander Gliboff has argued for a more significant influence—that Mendel's research project was an Ungerian one: find a mathematical law of evolution.[43] As Mendel's experiments were conceived and initiated in 1854, after his return to Brno from Vienna in 1853, both time and place make Unger a very probable influence for Mendel's experimental program.

There may be, however, an even more direct motivation for Mendel's experiments.

THE NATURE OF FERTILIZATION: MENDEL'S POSSIBLE CONTRIBUTION TO THE SCIENCE OF HIS TIME

According to Mendel's biographer Vitězlav Orel, Fenzl and Unger were in disagreement about the nature of fertilization. Fenzl argued that the embryo developed only from the pollen cell, whereas Unger argued that both parental plants contributed something to the new embryo.[44] Could it then be the case that as soon as he returned to Brno from Vienna, Mendel set this as a research question and attempted to resolve the dispute between his two professors in Vienna?

This is the suggestion made by botanist Rosalie Wunderlich,[45] given an intensive debate about the process of fertilization in higher plants that was taking place at the time that Mendel was a student in Vienna. Since 1830, there had been two opposing views on the topic. The Italian astronomer, mathematician, and microscope designer Giovanni Batista Amici, who had discovered the pollen tube in 1823, argued that a special cell (later called the "ovum") was present in the embryo sac before the arrival of the pollen tube. The latter had a stimulating effect on that cell, thus causing its development into an embryo and further to the whole plant. In contrast, the influential botanist Matthias Jacob Schleiden denied the existence of a cell in the embryo sac before the arrival of the pollen tube and argued that during the fertilization process the pollen tube grew into the embryo sac, which then served as the "incubator" of the embryo. As Schleiden was very influential as a botany professor and textbook author, many other botanists had come to accept his view. It was not until 1856–1857 that the physician and botanist Ludwig Radlkofer, a doctoral student of Schleiden who was studying fertilization and embryo development in higher plants, showed that Schleiden was wrong. But around the time that Mendel was a student in Vienna (1851–1853), the controversy between Schleiden and Amici was discussed. Unger agreed with Amici and cautiously expressed his view in his lectures and later in his publications. In contrast, Fenzl supported the views of Schleiden. We might infer whether one motivation for Mendel's experiments was to resolve this debate by looking again at what he wrote in his 1866 paper.

According to Wunderlich, the fact that Mendel was aware of this controversy is evident by what he wrote in the closing remarks of his paper:

> According to the opinion of famous physiologists, in phanerogams one germ and one pollen cell respectively unite to form a single cell that is able by absorption of matter and formation of new cells to develop itself further into a new autonomous organism.[46]

It is unclear who are the physiologists to whom Mendel referred in this sentence. However, Wunderlich believed that it was Unger, as he was the only one among other famous physiologists who by 1855 had expressed such a view. She argued that perhaps Mendel did not mention Unger by name because of the criticisms he had received from Brunner for holding views that would not align with those of a catholic friar like himself.[47] Whatever the case, what matters is that Mendel may have tried to provide decisive evidence for this. In a footnote he added to the previously quoted sentence, Mendel wrote:

> In Pisum it is surely beyond doubt that a complete union of the elements of both fertilization cells has to take place for the formation of the new embryo. How else would one want to explain that among the progeny of the hybrids both parental forms reemerge in the same quantity and with all their peculiarities? If the influence of the germ sack on the pollen cell was only external, if the role of a wet nurse was only assigned to the same, then the success of each artificial fertilization could be nothing else than that the developed hybrid resembled the pollen plant exclusively, or at least came to stand close to it. In no way has this been confirmed by the experiments so far. A thorough proof for the perfect union of the content of both cells presumably exists in the universally confirmed experience that it is immaterial for the conformation of the hybrid which of the parental forms was the seed or the pollen plant.[48]

So, what Mendel did, perhaps to resolve the debate between Unger and Fenzl, was to perform what is called reciprocal crosses: he used pollen from a plant with yellow seeds to fertilize a plant with green seeds, as well as pollen from a plant with green seeds to fertilize a plant with yellow seeds. If Fenzl was right and only the pollen contributed material to the embryo, then in the former cross all offspring should be yellow and in the latter all offspring should be green. If Unger was right and the contributions of both parents were equal, then the offspring in both crosses would be yellow, as this was the "dominant" character. Mendel's results clearly confirmed Unger's view.[49] As he wrote in his paper:

> It was shown throughout experiments, moreover, that it is completely irrelevant whether the dominating trait belongs to the seed—or the pollen plant; the hybrid form remains exactly the same in both cases. This interesting phenomenon is also emphasized by Gärtner, who notes that even the most skilled expert would be unable to distinguish in a hybrid which of the two conjoined species was the seed- or the pollen-plant.[50]

But according to Wunderlich, there is even more to this story.

Mendel had failed in his teacher examinations at the University of Vienna in 1850. He made a second attempt in 1856, although he had completed his studies already in

1853. Hugo Iltis, Mendel's first biographer, mentions a story that Mendel "had had a sharp difference of opinion with the examiner in botany, and had stubbornly maintained his own point of view". He also added that Nowotny, at that time a colleague of Mendel at the Brno "Modern School" believed that "this dispute with the examiner led Mendel to begin his experiments, whose origination certainly dates from very soon after the second defeat".[51] According to Wunderlich, that examiner could have only been Fenzl, as Unger may have not been allowed to serve as an examiner at that time due to Brunner's allegations and was also unlikely to have issues with Mendel. Even though one can assume that Mendel was well prepared for the botany exams, it is possible that he supported Unger's view of the equal parental hereditary contribution to the offspring, contrary to what Fenzl believed. Thus, either Mendel withdrew the exam, or Fenzl decided that he had failed.[52] In this view, Mendel's main motivation was to show that Fenzl was wrong, perhaps while being unaware that Schleiden's doctoral student Radlkofer had around the same time already shown that. It should be noted though that Iltis had expressed his "disbelief in the legend that a dispute between Mendel and the botanical examiner in 1856 gave the impetus to Mendel's investigations"[53]

Whether or not a dispute with Fenzl was the motivation for Mendel's experiments, there is no doubt that their results clearly provide evidence in support of the view of equal parental contribution during fertilization. Even though these results were eventually published ten years after the start of the experiments, at a time when there was no longer such a debate, this does not diminish the quality and the value of Mendel's experimental work toward establishing the equal hereditary contribution of both parental plants to the offspring. Most importantly, this can be seen as a direct contribution of Mendel to the science of his time, by answering questions about fertilization and hybridization.

The stereotypical representation of Mendel as a heroic, lonely pioneer who discovered the principles of heredity and established the appropriate experimental approaches for its study provides a distorted picture of how science in general is actually done. Science is a human activity, done within scientific communities, in particular (social, cultural, religious, political) contexts and is rarely—if ever—pursued by isolated individuals. Mendel was a man of his time and place, working on topics of interest to the sociocultural contexts in which he had lived and worked. His interests were broad, but his work was about the development of hybrids, having been influenced by the practical questions of interest in Brno as well as by the botanists the work of whom he had read or the lectures of whom he had attended. Contrary to the stereotypical account of Mendelian genetics, he did not try to develop a particulate theory of heredity.

Interestingly, at around the same time, and until the end of the 19th century there were several other naturalists who tried to develop theories of heredity based on postulated factors, elements, or particles.

NOTES

1 Knight (1824).
2 Roberts (1929), pp. 88–92.
3 Zirkle (1951).

4 Olby (1985), pp. 47–51.
5 Naudin (1864), p. 148 (author's translation).
6 Naudin (1864), p. 149 (author's translation).
7 Naudin (1864), p. 150 (author's translation; emphasis in the original)
8 Naudin (1864), p. 151.
9 Radick (2009).
10 Wood and Orel (2001), pp. 217–222.
11 Wood and Orel (2001), pp. 222–234.
12 Poczai (2022), pp. 72–78; Wood and Orel (2001), pp. 234–238.
13 Poczai (2022), pp. 96–102; Wood and Orel (2001), pp. 240–243, 270.
14 Wood and Orel (2001), pp. 252–258.
15 van Dijk et al. (2018), p. 353.
16 Mendel (2020), p. 110.
17 van Dijk et al. (2018), p. 349.
18 Müller-Wille and Orel (2007), p. 172.
19 Müller-Wille and Orel (2007), p. 175.
20 Mendel (2020), p. 27.
21 Olby (1985), pp. 3–4.
22 Mendel (2020), p. 27.
23 Müller-Wille and Orel (2007), p. 177. Table 1 of this paper presents occurrences of the term "Art" in Mendel's original paper, and some of its 20th-century English translations.
24 Olby (1985) pp. 4–5; Müller-Wille and Orel (2007), p. 182.
25 Olby (1985), p. 5, 11–13; Müller-Wille and Orel (2007), p. 183.
26 Olby (1985), p. 25.
27 Darwin Correspondence Project, "Letter no. 4612", accessed on 30 November 2022, www.darwinproject.ac.uk/letter/?docId=letters/DCP-LETT-4612.xml
28 Olby (1985), pp. 32–33.
29 Müller-Wille and Orel (2007), pp. 186–191.
30 Olby (1985), pp. 23–24, 26–31.
31 Shan (2020), pp. 20–21.
32 Rheinberger and Müller-Wille (2017), pp. 26–29.
33 Orel (1984) ch.3; Orel (1996) ch. 4.
34 Orel (1996), pp. 66–71.
35 Unger (1853).
36 Olby (1985), pp. 95–97.
37 Unger (1853), p. 107.
38 Unger (1853), p. 107.
39 Unger (1853), p. 108.
40 Olby (1967), 1985 pp. 199–205; Gliboff (1998).
41 Gliboff (1998).
42 Olby (1985), p. 207.
43 Gliboff (1999), p. 219.
44 Orel, (1984), p. 37.
45 Wunderlich (1983).
46 Mendel (2020), pp. 87–89.
47 Wunderlich 1983, p. 231
48 Mendel (2020), p. 87.
49 See also Fairbanks (2022), p. 64.
50 Mendel (2020), p. 35.
51 Iltis (2019/1932), p. 95.
52 Wunderlich (1983) pp. 232–233.
53 Iltis (2019/1932), p. 105.

3 Speculating about Heredity

THEORIES OF HEREDITY OF THE LATTER HALF OF THE 19TH CENTURY

The stereotypical narrative of Mendelian genetics gives the impression that Mendel was the only one who tried to develop a theory of heredity during the 19th century. In Chapter 1, I argued that Mendel was not trying to develop a theory of heredity and did not discover any laws of heredity. In Chapter 2, I argued that Mendel was interested in hybridization. However, questions about heredity were at the center of the biological thought of the time. The main reason was Charles Darwin's book *On the Origin of Species*, published in 1859, in which he wrote:

> The laws governing inheritance are quite unknown; no one can say why the same peculiarity in different individuals of the same species, and in individuals of different species, is sometimes inherited and sometimes not so; why the child often reverts in certain characters to its grandfather or grandmother or other much more remote ancestor; why a peculiarity is often transmitted from one sex to both sexes, or to one sex alone, more commonly but not exclusively to the like sex.[1]

As historians Hans-Jörg Rheinberger and Staffan Müller-Wille have explained, heredity became a central issue when due to Darwin's theory life forms stopped being considered as fixed by species boundaries. Darwin had proposed natural selection as the process of change, but it was also urgent to figure out the laws that maintained organisms in a stable state for at least some generations. In this way, once the idea of evolution was widely accepted, variation and heredity were seen as the two sides of the same coin.[2] There was thus a need for a process that could explain both why some traits varied across generations (variation, which is a prerequisite for natural selection) and why other traits remained stable across generations (heredity). As a result of this, several theories of heredity were proposed during the latter half of the 19th century: Charles Darwin's provisional hypothesis of pangenesis, Francis Galton's ancestral law of heredity, Ernst Haeckel's perigenesis of plastidules, G. Jaeger's physiological-chemical theory, William K. Brooks' law of heredity, Carl Nägeli's idioplasm theory, August Weismann's germ plasm theory, and Hugo de Vries' theory of intracellular pangenesis.[3] This resulted in an active community of scholars who essentially paved the way for the foundation of genetics, in parallel with other developments—particularly in cytology and microscopy—during that time. The fact that these scholars were influenced by one another is very important to emphasize as they were mutually influenced in various ways.[4]

These theories cannot be considered here in detail, but an overview of their main features is given in Table 3.1. In the present and the following chapter, we look at only

DOI: 10.1201/9781032449067-4

TABLE 3.1

Some Theories of Heredity of the Latter Half of the 19th Century

Author (Publication Year)	Hereditary Factors	Mechanism for Variation	Acquired Characters Inherited
Herbert Spencer (1864)	Physiological units contained in cells	Physiological units were remolded.	Yes
Charles Darwin (1868)	Gemmules thrown off from every unit of the body	Deficient amount of gemmules or modification of gemmules.	Yes
Ernst Haeckel (1876)	Plastidules with a frequency and amplitude of vibration	The frequency and amplitude of the vibration of plastidules could change due to the influence of external conditions.	Yes
Francis Galton (1876 and 1889)	Stirp (the sum total of developed and latent germs) existing in the body	Germs were not identical and might be modified if remained latent for long.	No
William Keith Brooks (1883)	Gemmules present in all cells	Gemmules, thrown off by affected cells, transmitted the change to the ovum.	No
Carl von Nägeli (1884)	Determinants contained in the idioplasm	Internal perfecting forces, external stimuli causing changes to the idioplasm, and sexual reproduction could give rise to intermediate characteristics.	No
Hugo de Vries (1889)	Pangens existing in cell nucleus into two groups (active and inactive)	Pangens of the same kind but of different origin might be activated, or pangens might be slightly dissimilar to the original ones after cell division.	No
August Weismann (1880s and 1892)	Biophors existing in the cell nucleus and forming units of higher order (determinants, ids, idants)	Modification of certain germ plasm determinants and differential combination of parental ids during fertilization.	No

some of these theories, which are related to the foundation of genetics. The theories considered in this chapter are those by Spencer and Darwin. Spencer was to first to write about heredity and to speculate about hereditary particles, but his writings are also an interesting example of how it is possible to anachronistically read a lot in a text of the past. Darwin's theory is also important because it was necessary for some of its preoccupations to be overcome for the path to genetics to open. This was in turn possible, thanks to the theories of Galton and Weismann, considered in Chapter 4, which mainly provided the new theoretical context in which Mendel's paper was reinterpreted in 1900. In Chapter 5 I also briefly consider the theories of Nägeli and de Vries.

The term "heredity" derives from the Latin *hereditas*, which means inheritance of succession. Therefore, the biological concept of heredity resulted from the metaphorical use of a juridical concept. This referred to the distribution of status, property, and other goods, according to a system of rules about how these should be passed on to other people once the proprietor passed away.[5] It is important to note that nowadays "heredity" is an exclusively biological concept, whereas "inheritance" is not as it is used in both biological and nonbiological contexts. The word "inheritance" is used colloquially to refer to whatever is passed on from parents to offspring, usually in the form of material or intellectual property. In the context of the life sciences, we refer to "genetic inheritance" to describe the process of transmission of genetic material across generations. Recently we have been talking about "epigenetic inheritance", or more broadly "non-genetic inheritance", to include processes that do not involve the information encoding DNA sequences. In contrast, "heredity" refers to the broader phenomenon of which these processes are part.

Let us now consider some of the first speculations about heredity.

SPENCER'S SPECULATIONS ABOUT HEREDITARY PARTICLES

The earliest references to heredity can be found in Herbert Spencer's (Figure 3.1) book *Principles of Biology*, which was published in 1863–1864, just a year before Mendel's paper was presented. Here is how he defined the concept of heredity:

> Understood in its entirety, the law [of hereditary transmission] is, that each plant or animal produces others of like kind with itself: the likeness of kind consisting, not so much in the repetition of individual traits, as in the assumption of the same general structure.[6]

He also noted that "A positive explanation of Heredity is not to be expected in the present state of Biology".[7] Spencer was also the first to propose a theory of heredity based on hereditary particles: the physiological units. In describing the regeneration of parts in various animals, he suggested that organisms were "made up of special units, in all of which there dwells the intrinsic aptitude to aggregate into the form of that species".[8] According to Spencer, groups of these units, provided that they were not differentiated, could re-arrange themselves and assume specific forms. He also suggested that these units could neither be the chemical compounds (which he called chemical units), nor the cells of the organisms (which he called morphological units).

From a photograph

HERBERT SPENCER

FIGURE 3.1 Herbert Spencer (1820–1903). Public Domain. Wikimedia Commons.

Rather, they should be units intermediate to those, and he called them physiological. Chemical units could combine into more complex units, which had the property of arranging themselves in a special form, due to their "organic polarity".[9] These polarities resulted from the special structures of the units, and, "by the mutual play of their polarities, they are compelled to take the form of the species to which they belong".[10]

Physiological units were the carriers of hereditary traits for Spencer. In his view, the available evidence showed that the reproductive cells were vehicles of these physiological units, in a state in which they could produce the structural arrangement of their species. He wrote:

> We must conclude that the likeness of any organism to either parent, is conveyed by the special tendencies of the physiological units derived from that parent. In the fertilized germ we have two groups of physiological units, slightly different in their structures.

During development, these two groups of units, which had similar polarities and tended to give rise to similar forms but had minor differences, produced together an organism of their species but also competed so that each would produce the traits of the parent from which it was derived. Consequently, the organism exhibited a mixture of parental traits.[11] However, if the physiological units tended to be arranged in a particular way, the characteristic of a species, then the existence of "parts slightly different from that of the species, implies physiological units slightly unlike those of the species".[12] In this way, for Spencer, physiological units were the factors both for heredity and for variation. This, in turn, allowed for the inheritance of acquired characters: "if the structure of this organism is modified by modified function, it will impress some corresponding modifications on the structure and polarities of its units". If there was no external influence, the units would produce a structure like the preexisting ones; if there was an external influence toward a new form, the units would be changed in order to be in harmony with this new form, and they would also build a similarly modified structure in the offspring.[13] Most importantly, given the number of physiological units contained in the reproductive cells, no two such cells from the same parent could be exactly alike. Consequently, the offspring of two parents would vary.[14]

In hindsight, there is a lot we can read in Spencer's speculations. Let's give it a try. Consider an interpretation of Spencer's speculations based on the knowledge that we currently have. There were hereditary units of intermediate size between molecules and cells, which he called physiological units (chromosomes and genes?). Those could multiply and produce similar new units (DNA replication?), although they might have slight differences under the influence of external circumstances, thus producing variability (mutation?). Every one of these units represented an entire specific character (genotype–phenotype relation?), and the respective units of different individuals were slightly different (alleles?). The two kinds of units of each parent excluded each other in normal fertilization (meiosis?), assuring uniformity within the individual. In other words, the child received parental units that made up its traits, and a strong resemblance to either one of the parents was due to the predominance of the respective physiological units (dominance?). My point here is that in hindsight, and based on the knowledge that we have today, we can read a lot of old writings, insofar as they fulfill our expectations. This is the kind of anachronistic reading that has been done for Mendel's paper and that had better be avoided.

In his analysis of the historiography related to Mendel, historian Staffan Müller-Wille has shown that their differences notwithstanding, many important historiographical analyses agree in one point: that there has been a divide between the historical works of geneticists describing in anachronistic terms the origins of their discipline, and those of professional historians that have attempted to debunk the myths that have emerged from the historical works of geneticists. Müller-Wille has shown that this divide has elements of myth itself, and has argued that the story is more complicated than that.[15] However, the important issue is that in order to understand what Mendel, Spencer, and any other was trying to do at the time, we need to consider their writings in their own historical, cultural, and social context. One thing is sure, though: even though August Weismann later acknowledged Spencer as the first in their generation to develop a theoretical explanation of heredity,[16] Herbert Spencer was not the first geneticist.

CHARLES DARWIN, "PANGENESIS", AND CONTINUOUS VARIATION

In 1868, Charles Darwin (Figure 3.2) presented his "Provisional Hypothesis of Pangenesis" in his book *Variation of Animals and Plants Under Domestication*.[17]

FIGURE 3.2 Charles Robert Darwin (1809–1882) around 1857. From Kostas Kampourakis, *Understanding Evolution*, 2nd edition © Kostas Kampourakis 2020, published by Cambridge University Press. Reproduced with permission.

He recognized that there were similarities between his own theory and that of Spencer; however, he considered his theory as a more advanced contribution.[18] Darwin attributed the concept of Pangenesis (Παγγένεσις), which literally means "origin from everywhere" (pan: all; genesis: origin), to Hippocrates.[19]

Even though Pangenesis was published in 1868, according to Darwin it had preoccupied him for many years. In an 1867 letter to Charles Lyell, he wrote:

> I have been particularly pleased that you have noticed Pangenesis. I do not know whether you ever had the feeling of having thought so much over a subject that you had lost all power of judging it. This is my case with Pangen: (which is 26 or 27 years old!) but I am inclined to think that if it be admitted as a probable hypothesis, it will be a somewhat important step in Biology.[20]

Why did he call it "provisional"? Perhaps because Thomas Henry Huxley, who read the first version of Pangenesis in 1865, had this to say: "But all I say is publish your views—not so much in the shape of formed conclusions—: as of hypothetical developments of the only clue at present accessible—and don't give the Philistines more chances of blaspheming than you can help".[21] Darwin also struggled for some time with the term itself. As he wrote to Huxley in 1867, he had thought of alternative terms but thought that "Pangenesis" worked best:

> Now I want to know whether I could not invent a better word.
> *Cyttarogenesis*, ie cell-genesis is more true & expressive but long.—
> *Atomogenesis*, sounds rather better, I think, but an "atom" is an object which cannot be divided; & term might refer to the origin of atom of inorganic matter.—
> I believe I like *pangenesis* best, though so indefinite; & though my wife says it sounds wicked like pantheism; but I am so familiar now with this word, that I cannot judge, I supplicate you to help me.[22]

Darwin had even asked his son George, at that time in Cambridge, to look for alternative terms.[23]

According to Darwin's hypothesis, all parts of the body participated in the formation of the offspring. The cells of the body, which multiplied by self-division and ultimately became converted into the various tissues, could throw off minute granules, called "gemmules", "which circulate freely throughout the system, and when supplied with proper nutriment multiply by self-division, subsequently becoming developed into cells like those from which they were derived". Gemmules were supposed to be transmitted from parents to offspring, being able either to develop in them or remain in a dormant state and develop many generations later. They also had a mutual affinity for each other that led to their aggregation and multiplication in the reproductive organs, thus forming the reproductive elements that were united in fertilization. Darwin concluded that "speaking strictly, it is not the reproductive elements, nor the buds, which generate new organisms, but the cells themselves throughout the body. These assumptions constitute the provisional hypothesis which I have called Pangenesis".[24]

A key issue in heredity was the reappearance in the offspring of traits that were not possessed by their parents but only by more remote ancestors—a phenomenon called reversion. Darwin assumed that the retention of undeveloped gemmules in the body was the only sufficient explanation for reversion: "the long-continued transmission of undeveloped gemmules explains many facts".[25] Darwin also believed that gemmules gave rise to new cells by uniting with nascent or partially developed cells, a process he compared to fertilization. "According to this view, the cells of the mother-plant may almost literally be said to be fertilised by the gemmules derived from the foreign pollen".[26]

Darwin also believed that variation was the outcome of the action of changed conditions during successive generations. More specifically variation resulted in two cases. On the one hand, it could be due to "the reproductive organs being injuriously affected by changed conditions; and in this case the gemmules derived from the various parts of the body are probably aggregated in an irregular manner, some superfluous and others deficient". On the other hand, variation could be due to "the direct action of changed conditions"; in such cases, the tissues of the body being directly affected by the new conditions could throw off modified gemmules, which were transmitted "with their newly acquired peculiarities to the offspring".[27]

In the first case, Darwin suggested that the partial deficiency of gemmules might cause considerable modifications in the body of the offspring. Thus, variation could emerge, and this could go on across the subsequent generations, gradually bringing about a significant modification. In the second case, under the conditions that caused the modification, gemmules continued multiplying until they became sufficiently numerous to replace the old unmodified ones. The modified gemmules would unite with the proper cells and would eventually give rise to a similar modified structure.[28] This was also the process through which the inheritance of acquired characters was possible.

It is important to note here that for Darwin the study of heredity required the study of development. It was during development that gemmules interacted with other preexisting cells, other gemmules, and external conditions of life to produce more cells, tissues, and organs. If, for some set of characters, the developmental path (from gemmule to character) of the offspring was similar to the path of the parent, inheritance had occurred; if the developmental path was different from the path of the parent, variation had occurred. In this sense, for Darwin, variation and heredity were opposed tendencies: heredity preserved parental characters, whereas variation pulled organisms away from them.[29] Darwin also considered hybridization and argued that the features of the offspring depended on which kind of gemmules would combine to form them. He assumed that hybrids should have both "hybridized gemmules" and pure, dormant gemmules derived from both parents. Depending on their combination, there could then be a complete or partial reversion of a character.[30] What Darwin called "hybridised gemmules" were gemmules thrown off by an intermediate-character part or organ grown from two different kinds of inherited gemmules that, as it were, collaborated in the making of that part.

In the *Origin of Species*, Darwin distinguished between two kinds of variation. He called the first one "individual differences", which were quite common.

> Again, we have many slight differences which may be called individual differences, such as are known frequently to appear in the offspring from the same parents, or which may be presumed to have thus arisen, from being frequently observed in the individuals of the same species inhabiting the same confined locality. No one supposes that all the individuals of the same species are cast in the very same mould. These individual differences are highly important for us, as they afford materials for natural selection to accumulate, in the same manner as man can accumulate in any given direction individual differences in his domesticated productions.[31]

So, according to Darwin, there were slight differences among individuals, which could be driven by natural selection to result in gradual change across generations. For instance, if some deer were slightly faster than others and thus had slightly higher chances of escaping predators than the "slower" ones, more and more of them would survive and reproduce, and so more and more of their descendants would be faster than their ancestors. As a result, across the generations, the average "fastness" of the population would increase because of natural selection.

The second kind of variation Darwin called "sports", which were rare.

A long list could easily be given of "sporting plants;" by this term gardeners mean a single bud or offset, which suddenly assumes a new and sometimes very different character from that of the rest of the plant. Such buds can be propagated by grafting, &c., and sometimes by seed. These "sports" are extremely rare under nature.[32]

Among these two, Darwin thought that individual differences were sufficient for natural selection: "A large amount of inheritable and diversified variability is favourable, but I believe mere individual differences suffice for the work".[33]

A key question was how species might maintain their character despite new variations. For Darwin it was sexual reproduction that did this. As he wrote in the *Origin of Species* "Intercrossing plays a very important part in nature in keeping the individuals of the same species, or of the same variety, true and uniform in character".[34] But it seems that Darwin was also aware that this maintenance of uniformity that intercrossing brought about could also counter the action of natural selection. He noted that although nature allowed for significant periods of time for natural selection to occur, time was not unlimited. As species were constantly striving to occupy specific niches within nature, if one failed to adapt and improve to the same extent as its competitors, it could eventually die out. In the case of deliberate human selection, breeders could choose specific traits for a particular purpose, and the process could be halted by unrestricted interbreeding. Darwin thought that for natural selection to operate, it had to take place within a restricted area so that individuals with a specific kind of favorable variation could mate with one another and thus produce offspring carrying this favorable variation. In this way, slow and gradual modification by means of natural selection could occur. This was in some sense what breeders were doing as they did not let all individuals in a group mate randomly but rather selected which ones to mate. However, if mating occurred randomly over larger areas, then this could act against natural selection.[35]

This was exactly the kind of criticism that Scottish engineer Fleeming Jenkin made in an anonymous review of the *Origin of Species* that appeared in 1867 in the *North British Review*. He addressed the aforementioned Darwin's concern, using a vivid, and full of racial prejudice, illustrative example. Jenkin proposed an imaginative scenario in which "a white man" got stranded on an island inhabited by "negroes" and eventually managed to establish friendly relations with a powerful tribe. (Jenkin noted though that the superiority is not inherently tied to the white color but only used whiteness to illustrate how traits associated with a particular individual within a larger population gradually diminish over time.) This man possessed the physical strength and abilities typically associated with the "dominant white race", whereas the island's food and climate were good for him. One could thus assume, according to Jenkin, that our hero had every advantage a white person might have had over the native inhabitants. One could also assume that due to his superior chances of survival in the struggle for existence, his lifespan would be longer compared to that of the natives. Jenkin suggested that our hero might thus become a king and eliminate many members of the native population and father numerous children with his many wives, while many natives would eventually die childless. However, this would not necessarily entail that the island's population would become predominantly white after a certain number of generations. Even over multiple generations, his influence alone would not be sufficient "to turn his subjects' descendants white".[36] This is often described as the "swamping" effect. Simply put, whatever superior traits one or a few individuals may have, as they are just a very small proportion of the population, it is more likely for them to mate with individuals with lower qualities. Even if their offspring had relatively higher qualities than the others in the population, it would similarly be more likely for them to mate with individuals with lower qualities. If this continued across generations, the superior traits would be "swamped" by being diluted into a much larger pool of regular traits.

Darwin took this criticism seriously. In a letter to Wallace on January 22, 1869, Darwin wrote:

> I have been interrupted in my regular work in preparing a new edit of the Origin, which has cost me much labour & which I hope I have considerably improved in two or three important points. I always thought individual differences more important than single variations, but now I have come to the conclusion that they are of paramount importance, & in this I believe I agree with you. Fleming Jenkyn's arguments have convinced me.[37]

And then again on February 2, 1869:

> F. Jenkins argued in N. Brit. R. against single variations ever being perpetuated & has convinced me, though not in quite so broad a manner as here put.— I always thought individual differences more important, but I was blind & thought that single variations might be preserved much oftener than I now see is possible or probable.[38]

Darwin took into account Jenkin's arguments in the 5th edition of the *Origin of Species*, published in 1869.

It has been argued that it was not Jenkin's point about swamping but rather about the shortness of geological time that was more influential for Darwin.[39] Most importantly, an error in Jenkins' calculations was pointed out by A.S. Davis in 1871. Jenkin had assumed that each parent pair produced one offspring, but for the population size to be constant he should have assumed that each parent pair had produced two offspring. Davis also pointed out that if a favorable variation occurred once, it could hardly be sufficient to bring about any change. However, if this happened independently in different generations, although not more than once in each one of them, it could bring about significant change.[40] In other words, if the same variation emerged several times across numerous generations, it would be more difficult for it to be "swamped". Therefore, the argument about the swamping effect that Jenkins had made was rather exaggerated. Darwin's account of natural selection wasn't anything like as vulnerable to swamping objections as was subsequently suggested, though his critique did prompt Darwin to clarify his views on the kinds of variation that mattered for natural selection along lines that Darwin valued.

This notwithstanding, what I want to point out here is that Darwin was aware before Jenkin's review that continuous variation and gradual evolution by natural selection were possible only under particular conditions. Concerns about Darwin's ideas regarding the nature of variation and how evolution proceeds had already been raised by Thomas Henry Huxley in 1859. As soon as he read the *Origin of Species*, Huxley wrote a letter to Darwin stating:

> The only objections that have occurred to me are 1st that you have loaded yourself with an unnecessary difficulty in adopting 'Natura non facit saltum' so unreservedly. I believe she does make small jumps—and 2nd. it is not clear to me why if external physical conditions are of so little moment as you suppose variation should occur at all.[41]

The unnecessary difficulty was that if evolution was gradual, then one should have expected to see all kinds of intermediate forms—particularly in the fossil record. In contrast, if small jumps—"saltations" as Huxley called them—were possible, then there would be no such need for intermediate forms.[42]

"PANGENESIS" PUT TO THE TEST

Between 1869 and 1871, Francis Galton, Darwin's half-first cousin (more on whom in the next chapter), tried to test experimentally the hypothesis of Pangenesis by performing transfusions of blood between different strains of rabbits. Darwin was aware of the experiments, was constantly kept informed about the results, and did not discourage them at all, whereas Galton had hoped that they would indeed provide evidential support for Pangenesis.[43] When Galton obtained a rabbit with a white foot, a trait that did not exist in the parents, he wrote to Darwin:

> My dear Darwin
>   Good rabbit news.! One of the latest litters has a white forefoot. It was born April 23rd. but as we do not disturb the young, the forefoot was not observed till to-day. the little things had huddled together shewing only their backs & heads and the foot was never suspected. The mother was injected from a grey and white and the father from a black and white.
>   This, recollect, is from a transfusion of only 1/8th part of alien blood in each parent; now, after many unsuccessful experiments, I have greatly improved the method of operation and am beginning on the other jugulars of my stock. Yesterday I operated on 2 who are doing well to-day & who now have 1/3rd. alien blood in their veins. On Saturday I hope for still greater success. and shall go on at any waste of rabbit life, until I get at least 1/2 alien blood.
>   The experiment is not fair to Pangenesis until I do.[44]

In a letter Darwin's wife Emma sent to their daughter Henrietta, she wrote:

> F. Galton's experiments about rabbits (viz. injecting black rabbit's blood into grey and *vice versa*) are failing, which is a dreadful disappointment to them both. F. Galton said he was quite sick with anxiety till the rabbits' *accouchements* were over, and now one naughty creature ate up her infants and the other has perfectly commonplace ones. He wishes this experiment to be kept quite secret as he means to go on, and he thinks he shall be so laughed at, so don't mention.[45]

However, Galton's official report of the experiments, published in the *Proceedings of the Royal Society* in 1871, was negative:

From this large stock I have bred eighty-eight rabbits in thirteen litters, and in no single case has there been any evidence of alteration of breed. There has been one instance of a sandy Himalaya; but the owner of this breed assures me they are liable to throw them, and, as a matter of fact, as I have already stated, one of the does he sent me, did litter and throw one a few days after she reached me. The conclusion from this large series of experiments is not to be avoided, that the doctrine of Pangenesis, pure and simple, as I have interpreted it, is incorrect.[46]

Galton's experiments were based upon the following assumption: Darwin had written about "minute granules . . . that circulate freely throughout the system", "circulation of fluids" as well as that "the gemmules in each organism must be thoroughly diffused". To Galton the words "circulate" and "diffused" implied a connection to the circulation of the blood. He thus transfused the blood of other species of rabbit to the blood vessels of male and female silver-gray rabbits, which were afterward bred. In some cases, he made so many transfusions to particular rabbits that he estimated that about half of their blood was "alienized".[47] If Pangenesis was correct, there would have been a change in the color of the successive generations, due to the gemmules transferred during the transfusions. However, after repeating the procedure for three generations, Galton could find no sign of any deterioration in the purity of the silver-gray breed. These negative results seemed to contradict the hypothesis of Pangenesis. However, Darwin objected to the assumption of blood circulation as it was "no necessary part of the hypothesis". He noted that he did not write anything about blood in his book and that he also referred to heredity in protozoa and plants that do not have blood. His assumption was that gemmules could be transferred from one cell to another and across tissues independently of the presence of vessels of any kind. What Galton should have done, according to Darwin, would be to show that gemmules were indeed found in the blood of animals. Galton replied that he was misled by the inappropriate use of the words "circulate", "freely", and "diffuse" by Darwin.[48] It seems that Darwin was influenced by this criticism; when he wrote about gemmules in the second edition of the *Variation of Animals and Plants Under Domestication*, he wrote that they "are dispersed throughout the whole system"[49] instead of what he had written in the first edition, that they "circulate freely throughout the system".[50]

Darwin initially believed that the views of Rudolf Virchow on the cellular basis of life supported his own views of the independent life of each minute element of the body, which is the key concept of Pangenesis. However, Pangenesis was thus inconsistent with the claim that all cells must be derived from other cells. Therefore, the theory, as presented in the first edition, was more consistent with the old cell theories rather than with the cytology of the 1860s.[51] However, in the second edition of the book and his correspondence, Darwin clarified that he conceived of the gemmules as leaving one specialized cell, and upon entering a less specialized cell (like a stem cell), the latter was (or might be) transformed into a mature cell type of the former donor cell.[52]

Darwin thought that Pangenesis was plausible because it seemed to explain a wide range of phenomena, as natural selection did. In his autobiography, he wrote about his "well abused hypothesis of Pangenesis".

An unverified hypothesis is of little or no value. But if any one should hereafter be led to make observations by which some such hypothesis could be established, I shall have done good service, as an astonishing number of isolated facts can thus be connected together & rendered intelligible.[53]

Indeed, Pangenesis proved very important. It was a theory of heredity with a purely material basis and, without any need for vitalism, in contrast to what many naturalists thought at the time.[54] Furthermore, it did point to the problems that a theory of heredity had to resolve. And, as Galton later pointed out, he did find there a key element for his subsequent work: "the idea, though not the phrase of particulate inheritance, is borrowed from Darwin's Provisional Hypothesis of Pangenesis".[55]

Galton initiated a shift toward a better understanding of heredity, which followed by Weismann's contribution did in fact pave the way for the science of genetics.

NOTES

1 Darwin (1859), p. 13.
2 Rheinberger and Müller-Wille (2016), pp. 146–147.
3 See Robinson (1979) for a concise overview.
4 Kampourakis (2013).
5 Müller-Wille and Rheinberger (2012), pp. 5–6.
6 Spencer (1864), p. 238.
7 Spencer (1864), p. 253.
8 Spencer (1864), p. 181.
9 Spencer (1864), pp. 181–182.
10 Spencer (1864), p. 253.
11 Spencer (1864), p. 254.
12 Spencer (1864), pp. 254–255.
13 Spencer (1864), pp. 256
14 Spencer (1864), pp. 266.
15 Müller-Wille (2021).
16 Weismann (1893), p. 1.
17 Darwin, 1868, pp. 357–404.
18 Darwin 1868, pp. 374–375.
19 Darwin (1868), p. 375.
20 Darwin Correspondence Project, "Letter no. 5612," accessed on 11 March 2023, www.darwinproject.ac.uk/letter/?docId=letters/DCP-LETT-5612.xml
21 Darwin Correspondence Project, "Letter no. 4875," accessed on 11 March 2023, www.darwinproject.ac.uk/letter/?docId=letters/DCP-LETT-4875.xml
22 Darwin Correspondence Project, "Letter no. 5568," accessed on 11 March 2023, www.darwinproject.ac.uk/letter/?docId=letters/DCP-LETT-5568.xml
23 Darwin Correspondence Project, "Letter no. 5561," accessed on 11 March 2023, www.darwinproject.ac.uk/letter/?docId=letters/DCP-LETT-5561.xml
24 Darwin (1868), p. 374.
25 Darwin (1868), p. 378.
26 Darwin (1868), p. 388.
27 Darwin (1868), pp. 394–395.
28 Darwin (1868), pp. 394–395.
29 Winther (2000).
30 Darwin (1868) pp. 400–401.

31 Darwin (1859), p. 45.
32 Darwin (1859), pp. 9–10.
33 Darwin (1859), p. 102
34 Darwin (1859), p. 103
35 Darwin (1859), pp. 102–103.
36 Jenkin (1867), pp. 289–290.
37 Darwin Correspondence Project, "Letter no. 6567," accessed on 10 March 2023, www.darwinproject.ac.uk/letter/?docId=letters/DCP-LETT-6567.xml
38 Darwin Correspondence Project, "Letter no. 6591," accessed on 11 March 2023, www.darwinproject.ac.uk/letter/?docId=letters/DCP-LETT-6591.xml
39 Morris (1994).
40 Davis (1871), p. 161.
41 Darwin Correspondence Project, "Letter no. 2544," accessed on 11 March 2023, www.darwinproject.ac.uk/letter/?docId=letters/DCP-LETT-2544.xml
42 Provine (2001), pp. 10–13.
43 Bulmer (2003), pp. 116–119.
44 Darwin Correspondence Project, "Letter no. 7185," accessed on 10 February 2023, www.darwinproject.ac.uk/letter/?docId=letters/DCP-LETT-7185.xml
45 Litchfield (1915), p. 197.
46 Galton (1871a), pp. 403–404.
47 Galton (1871a), p. 400.
48 Darwin (1871); Galton (1871b).
49 Darwin (1875), p. 370.
50 Darwin (1868), p. 374.
51 Hodge (1985), p. 227.
52 Reynolds (2018).
53 Darwin (1876–1882), pp. 104–105.
54 Robinson (1979), pp. 18–19.
55 Galton (1889), p. 193.

4 The Path to Genetics

FRANCIS GALTON AND DISCONTINUOUS VARIATION

Francis Galton (Figure 4.1) made many important contributions to geography, meteorology, fingerprinting identification, anthropometry, psychology, and photography. However, what he is mostly remembered for, which is also the reason that his name makes people dismayed, is as the founder of eugenics (see Chapter 8). Galton was already interested in heredity before Pangenesis was published. His first paper on the topic, titled "Hereditary talent and character", was published in 1865—the same year that Mendel presented his findings, for those interested in coincidences (another is that they were both born in the same year, 1822). At the beginning of that paper, Galton stated that "we must ever bear in mind our ignorance of the laws which govern the inheritance even of physical features".[1] To investigate the hereditary transmission of talent, he collected information about intellectual achievements from several genealogies and concluded that

> intellectual capacity is so largely transmitted by descent that, out of every hundred sons of men distinguished in the open professions, no less than eight are found to have rivalled their fathers in eminence. It must be recollected that success of this kind implies the simultaneous inheritance of many points of character, in addition to mere intellectual capacity.[2]

In the same paper, Galton also rejected the idea of the inheritance of acquired characters. He asked whether the "virtuous habits" of parents result in their children being born with "more virtuous dispositions" or whether we are just "passive transmitters of a nature we have received", which we cannot modify, concluding that "There are but a few instances in which habit ever seems to be inherited".[3] Another key principle that he mentioned in that paper was that "the transmission of talent is as much through the side of the mother as through that of the father".[4] Biparental inheritance was widely accepted by that time, and Galton practically suggested that heredity was diluted across generations because each person received one-half of the hereditary constitution of each of their parents, one-quarter of that of each of their grandparents, and so on. Thus, a reason for sons being less distinguished than their fathers was, according to Galton, that they had received only half of their constitution from them and the other half from their mother.[5] Another reason for this was reversion: the reappearance in individuals of traits exhibited by remote ancestors but not their own parents. No matter how talented the parents might be, it would be "absurd to expect their children to invariably equal them in their natural endowments".[6] The tendency for traits of remote ancestors to reappear was also something that Darwin himself had considered important.

A key idea proposed in that 1865 paper was that offspring descended from the embryo of their parents.

DOI: 10.1201/9781032449067-5

FIGURE 4.1   Francis Galton (1822–1911) around 1864. Galton Papers (GALTON/1/1/12/3/4), UCL Special Collections, UCL Archives, London.

> We shall therefore take an approximately correct view of the origin of our life, if we consider our own embryos to have sprung immediately from those embryos whence our parents were developed, and these from the embryos of their parents, and so on for ever.[7]

This idea was more fully elaborated in a paper published in 1872, in which Galton set the basis for his theory of heredity, inferring again from the fact of reversion the presence of elements that remained latent in individuals, while others—the patent ones—were expressed. Galton in fact arrived at what he thought of as a general, and safe, conclusion that the patent elements contributed a lot less than the latent elements. The reason for this, he argued, was the poor transmission of what he described as "personal elements", which do not have any latent equivalents. Galton concluded from the fact that acquired characters were rarely, if at all, inherited that these personal elements were only partially inherited.[8] He concluded that parents were "very indirectly and only partially related to their own children" through two lines of connection: an indirect one from the same source, which was the important one, and a direct one between them which was less important.[9]

The key idea was that what mattered for heredity was not the elements that parents themselves expressed but rather those that they carried. Essentially, it was Galton's acceptance of Darwin's idea that those gemmules that develop into cells are not transmitted to the next generation, and Galton's rejection of Darwin's idea that cells can throw off new gemmules, that led Galton to the conclusion that only latent elements were inherited. Offspring and parents were not directly connected but only indirectly through their "primary elements" as Galton put it. In sharp contrast with Darwin, who thought that an individual's personal qualities were transmitted to their offspring, Galton thought that they were not; what was inherited, rather, were those "primary elements", and what connected parents to offspring was that their elements were derived from a common source.[10] Needless to say that this theory could not really explain much about the similarities between parents and offspring, which was what probably Galton was most interested in. How could similarities between parent and offspring be explained if the respective patent elements were not transmitted? This could be possible only by assuming a correlation between those particular patent elements and corresponding latent elements that were inherited by the offspring. But Galton offered no explanation for this.[11]

In 1876, Galton published a fuller account of his theory of heredity. He suggested that a complete theory of heredity must account for both the "inborn or congenital peculiarities" that were the same at least in one ancestor and those that "were not congenital in the ancestors" but were acquired due to some changes in the conditions of life. Consequently, a complete theory of heredity should be divided into two corresponding parts of which "the first stands by itself, the second is supplementary or subordinate to it".[12] For Galton Pangenesis could be accepted only as a theory of the latter kind, even though he wrote that

> No theory of heredity has been enunciated with more clearness and fulness than that of Mr. Darwin's Pangenesis, and the preparatory statement to that theory contains the most elaborate epitome that exists, of the many varieties of facts for which a complete theory of heredity must account.[13]

Galton accepted that the existence of organic units should lie at the foundation of the science of heredity, and he defined those as "stirp". This term was derived from the Latin stirpes, meaning a root, "to express the sum-total of the germs, gemmules, or whatever they may be called, which are to be found, according to every theory of organic units, in the newly fertilised ovum".[14] Galton then described four postulates necessarily implied by any hypothesis of organic units, which were also included in Pangenesis: (1) that each independent unit of which the body consisted had a separate origin or germ; (2) that the stirp contained more germs than those of which the units of the body were derived; (3) that these other germs that remained undeveloped retained their vitality, propagated themselves, and formed the stirps of the offspring; and (4) that the whole organization depended on the mutual affinities and repulsions of the separate germs. He noted that "Proofs of the reasonableness of these postulates are especially to be found in the arguments of Mr. Darwin".[15] This is again evidence of the influence of Pangenesis at least on theoretical grounds.

Galton then described how reproduction and development occurred, and he summarized the main conclusions about the role of the stirp in these processes. There was a constant competition among certain germs for each unit of structure, and some managed to become developed into cells, whereas others did not and rested in a dormant state. The dominant germs eventually participated in the development of the individual, whereas the latent germs formed the sexual elements from which the offspring emerged.[16] For Galton individual variation depended upon two factors: "the variability of the germ and of its progeny" and "all kinds of external circumstances" that determined "which out of many competing germs, of nearly equal suitability, shall be the one that becomes developed". There was a strong tendency of like producing like, but there were exceptions too. Eventually, it was the competition among germs that determined the final outcome. These were of different quality and could also be modified while being in the dormant state, thus potentially leading to a marked individual variation.[17] This was a critical difference from Darwin's views. External conditions did not modify the parts of the body that in consequence released germs, modified in the same manner. What they did was to favor one of the qualitatively different competing germs (Galton called them dominant) to become developed. Only in the dormant state was it possible for a germ to be modified, and this could happen during many successive generations.

Galton did not completely rule out the idea that characters acquired during life by a parent could be transmitted to their offspring. If Darwin's hypothesis was correct, according to Galton, a change in the body should have affected the reproductive elements.[18] But he objected to this idea, on the grounds that "There is not a shadow of proof that the adaptivity of a race to changed conditions, *affecting all parts of the body alike*, is due to the reaction of changed personal structure upon the sexual elements". He also added that "the certainty of the non-inheritance of mutilations in a vast number of cases . . . is so over-powering".[19] Galton thus concluded that acquired modifications were barely, if at all, inherited.[20] Despite this, his explanations about the similarities between relatives remained confused. Regarding siblings, he wrote that they had in principle the same elements, and any differences between them were due to variability emerging during development. Monozygotic twins were very similar because there was no developmental variability between them.[21]

An interesting summary of Galton's views of heredity was made in a letter he wrote to Darwin on 19 December 1875. In a letter written on the previous day, Darwin had challenged Galton's theory of heredity by asking him to explain how plant hybrids intermediate in character between their parents produced sexual elements that exactly reproduced that intermediate character.[22] Galton replied, using an example of an animal in which the hybrid between white and black forms was gray, an exact intermediate. He then provided diagrams showing a white and a black form, consisting of white and black cells, respectively. He argued that in the hybrid either some cells would be black and others would be white in equal proportions, resulting in an overall uniform gray character, which he dubbed (1), or all cells would each be uniformly gray, which he dubbed as (2). The structural units in (1) would be cells, whereas in (2) would be molecules. Galton continued:

> The larger the number of gemmules in each organic molecule, the more *uniform* will the tint of grayish be in the different units of structure. It has been an old idea of mine, not yet discarded & not yet worked out, that the number of units in each molecule may admit of being discovered by noting the relative number of cases of each grade of deviation from the mean grayness. If there were 2 gemmules only, each of which might be either white or black, then in a large number of cases one quarter would always be quite white, one quarter quite black, & one half would be gray.[23]

Sounds familiar? These are like Mendel's ratios. Crossing white with black resulted in 1/4 black, 2/4 gray, and 1/4 white. But, in contrast to Mendel, Galton clearly and explicitly postulated the existence of hereditary particles (which he called "structural units").

Subsequently, Galton switched to developing a statistical, rather than a physiological, theory of heredity. His results were summarized in his book *Natural Inheritance*. This work resulted in a modification of his earlier theory and, in particular, of the role or patent and latent elements. However, he did not explicitly acknowledge that change, nor was he consistent in that respect.[24] The most significant change was that Galton no longer thought that latent elements were more likely to be inherited than patent elements. Rather, he assumed that patent and latent elements were equally likely to be transmitted to the offspring.[25] In this work, Galton also proposed the concept of "sport" (already used by Darwin, see Chapter 3), which referred to the emergence of a new "type" through marked changes.[26] This was based on a model of a polygon, the edges of which were uneven in length (Figure 4.2). Here is how it worked: at first the polygon has a stable, symmetric configuration. When it is pushed with sufficient force it will rest on an asymmetric, unstable edge. From that position it may get back to its original stable position or rest on the next edge, which is another stable, symmetric position. If the latter happens, then the polygon will have achieved an entirely new state of stability. Galton thus proposed a mechanism for discontinuous variation, illustrating how certain conditions might co-exist: "(1) variability within narrow limits without prejudice to the purity of the breed; (2) Partly stable sub-types; (3) Tendency, when much disturbed, to revert from a subtype to an earlier form; (4) Occasional sports which may give rise to new types".[27]

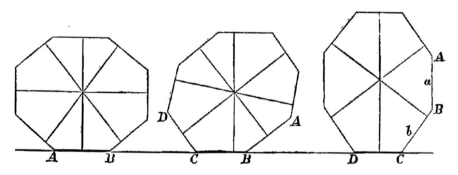

FIGURE 4.2   Galton's model of a polygon with slightly asymmetric surfaces as an illustration of his hypothesis of organic stability. In the left diagram the polygon rests on edge A–B. This is a stable, symmetric configuration. Only when it is pushed with sufficient force it will tumble on the next edge C–B, an asymmetric, unstable edge. A slight push from the front will bring the polygon back to rest in its original stable position, whereas a heavy push from behind will bring the polygon to rest on edge D–C, which is another stable, symmetric position. The polygon has then achieved an "entirely new system of stability"—in Galton's words, "a sport comes suddenly into existence". From Galton (1889), p. 27.

Here is Galton's decisive breaking with Darwin. Evolution was not necessarily gradual, because variation was not necessarily continuous. In *Natural Inheritance*, he began a section titled "Evolution Not by Minute Steps Only" by writing:

> The theory of Natural Selection might dispense with a restriction, for which it is difficult to see either the need or the justification, namely, that the course of evolution always proceeds by steps that are severally minute, and that become effective only through accumulation. That the steps may be small and that they must be small are very different views; it is only to the latter that I object, and only when the indefinite word "small" is used in the sense of "barely discernible", or as small compared with such large sports as are known to have been the origins of new races.[28]

Galton was a strong proponent of the idea that variation is discontinuous because of the phenomenon of regression to the mean, which was an idea very close to Jenkin's "swamping" argument (see Chapter 3). Galton wrote in *Natural Inheritance*, in a section titled "Regression":

> However paradoxical it may appear at first sight, it is theoretically a necessary fact, and one that is clearly confirmed by observation, that the Stature of the adult offspring must, on the whole, be more mediocre than the stature of their Parents; that is to say, more near to the M of the general Population.[29]

This is regression. Galton actually estimated the respective amount as having the value of two-thirds, on average.[30] What Galton did was to calculate the average height of the two parents (he called it "mid-parental") and then compare this to the median height of their children (he called it "mid-filial"). He then plotted these median values against the mid-parental height, and what he found was a deviation between the two lines—something not to be expected if children were as tall as their parents. In contrast, the results showed regression toward mediocrity. When mid-parents were taller than mediocrity, their children tended to be shorter than them.[31]

One central idea in Galton's theory was that half of the hereditary particles of each individual were passed on to each offspring, and this was a consequence of having two parents who contributed equally to the heritage of their offspring. Galton wrote in 1876: "There is yet another advantage in double parentage, namely, that as the stirp whence the child sprang, can be only half the size of the combined stirps of his two parents, it follows that one half of his possible heritage must have been suppressed".[32] And again in 1889: "It is not possible that more than one half of the varieties and number of each of the parental elements, latent or personal, can on the average subsist in the offspring".[33] However, Galton never made the additional step toward explaining the variability of siblings on the basis of the segregation of hereditary particles—something that was clearly stated for the first time by August Weismann in 1887 (in German, the English translation of that essay was published in 1889). Whereas Galton correctly realized that it was necessary for the number of hereditary particles to become half, so that both parents contributed equally, he did not realize that this could also explain the variability in offspring.

In 1897 Galton proposed his "Ancestral Law of Heredity":

> It is that the two parents contribute between them on the average one-half, or (0.5) of the total heritage of the offspring; the four grandparents, one-quarter, or (0.5)^2; the eight great-grandparents, one eighth, or (0.5)^3, and so on. Thus the sum of the ancestral contributions is expressed by the series {(0.5)+(0.5)^2+(0.5)^3, &c.}, which, being equal to 1, accounts for the whole heritage.[34]

This means that, on average, one-half of the hereditary particles of a person came from their parents, and the other half came from their more remote ancestors. Galton noted that the more remote one's ancestors were, the less they had contributed to a person's hereditary qualities, compared to their more recent ancestors. This ancestral contribution was possible because for Galton the main, if not the only, line of descent run from germ to germ and not from person to person. He also pointed out that the law was in agreement with the halving of the chromosome numbers during the reduction division (meiosis—see the next section). In 1898, he also had an illustration of this law published (Figure 4.3). However, in Galton's formulation of this law, there was a confusion due to the lack of clear distinction between its two possible interpretations: (1) as a representation of the contributions of different ancestors to an individual, and (2) as a regression formula for predicting the value of a trait from those of the ancestors. The problems in the initial proposition of this law were later addressed by Karl Pearson, who developed the theory of multiple regression.[35]

There are two important concepts in Galton's conceptualization of heredity, both of which sound familiar in the context of Mendelian genetics:

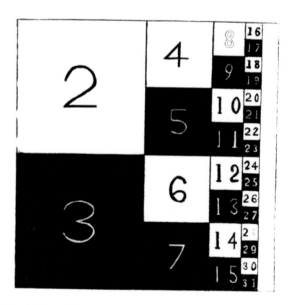

FIGURE 4.3   Galton's Law of Ancestral Heredity in a diagram devised by Mr. A.J. Meston, of Allen Farm, Pittsburgh, Massachusetts, USA, and communicated by him to the *Horseman*, the leading American newspaper on horse-breeding. Reproduced with permission from Galton (1898).

1. That there was a germline, distinct from the substance of the main body, from which offspring emerged.
2. That variation was not necessarily continuous and could be discontinuous.

Whereas Galton was first and foremost a theoretician, it was August Weismann who made Galton's ideas biologically meaningful. The advances in cytology of the time were crucial for this.

THE ADVANCES IN CYTOLOGY OF THE 1870s

In 1873, four zoologists working independently observed the complex changes that the cell nucleus underwent during cell division. These were Anton Schneider, working with the flatworm *Mesostomum*, who used acetic acid to make the disappearing outline of the nucleus evident and thus managed to observe the transformation of the substance of the nucleus to cords arranged in line with two poles; Hermann Fol, working with the hydroid *Geryonia*, who found that the figures that grew out of the poles of the disappearing nucleus became the centers of the formation of two new nuclei; Otto Bütschli working with the nematode *Rhabditis dolichura*, who observed the fusion of the two "pronuclei" during fertilization followed by the appearance of "knobs" (probably the centromeres in retrospect) that separated to the opposite poles; and Walther Flemming, working with freshwater mussel, who observed rays with granules radiating toward two poles where the two new nuclei would then form. Soon others joined these studies, and between 1875 and 1877 important advances were made. Leopold Auerbach's work with the nematodes *Ascaris* and *Strongylus* confirmed Bütschli's observations about the union of pronuclei during fertilization. He also argued that cell nuclei could become visible when tissues were treated with the appropriate reagents and that inside them one could find many small particles of variable sizes. Oscar Hertwig provided a detailed description of fertilization in the sea urchins *Toxopneustes lividus*, suggesting that the ovum and sperm pronuclei fused and produced a new nucleus. Importantly, he also addressed the question that had divided zoologists and botanists at the time about whether the nucleus was dissolved before cell division or whether it persisted through division. Hertwig suggested that the nucleus underwent a metamorphosis during cell division, which could account for its apparent dissolution, with the various parts of it practically maintained. Another contribution was made by Waclaw Mayzel, who managed to observe three distinct forms of the nucleus during cell division. Eduard Strasburger established that much of what zoologists had observed happening in animal cells was also happening in plant cells. Finally, Richard Hertwig, Oscar's brother, tried to establish the nature of the nucleus across all different kinds of organisms, against the lack of a generalized understanding of it.[36]

Despite all these people involved in the study of cell nucleus, it may be worth highlighting the contributions of Walther Flemming (Figure 4.4) between 1877 and 1882. He used the epithelial lining of the bladder of the salamander *Salamandra maculate*, which contained large cells with nuclei that were easily observed. By 1878, Flemming had found evidence supporting the conclusion that the substance inside the nucleus persisted during the resting stage of the cell. He thus questioned the

FIGURE 4.4   Walther Flemming (1843–1905), in the 1870s. Public Domain. Wikimedia Commons.

prevailing assumption that the nucleus divided through fission at the same time that the protoplasm was also divided, which was described as "direct cell division". He proposed a new alternative that he described as "indirect cell division" because he had observed that the substance of the nucleus underwent a transformation before fission occurred. Flemming was not the first to make such observations; however, it was he who showed that the "threads" observed could split longitudinally during the division. His findings were presented in detail in his 1882 book *Zellsubstanz, Kern und Zelltheilung* (*Cell Substance, Nucleus and Cell Division*). Flemming had a talent for drawing, and his 1882 book is illustrated with fine and detailed drawings that represent the division of the nucleus (Figure 4.5). His methods allowed the distinction between a material in the nucleus that could be stained, which he called "chromatin", and other structures that could not be stained, which he called "achromatin". Flemming described the division of the nucleus as we know it today, making a distinction between a "progressive" phase, during which the threads appeared and ended up being arranged in the center of the cell, and a "regressive" phase, during which the threads were separated into two groups that ended up in each of the new nuclei (Figure 4.5). Flemming called this process "Karyomitosis" (meaning thread-like metamorphosis of the nucleus) and the arrangements of the nuclear threads "Mitosen"—what we today call mitosis.[37]

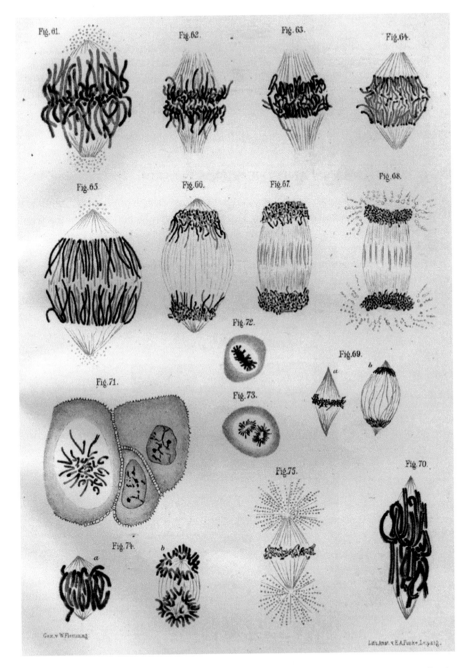

FIGURE 4.5 Illustrations of the division of the nucleus, which Flemming called Karyomitosis, from his 1882 book *Zellsubstanz, Kern und Zelltheilung* (*Cell Substance, Nucleus and Cell Division*). Reproduced with permission from Paweletz (2001).

In his 1882 book, Flemming concluded a section presenting the morphological details of the ovum with the following expectation:

> Should the substance of the egg, however, have a structure, then this and the creation of threads in particular areas of the cell body can be different, so also therein can a basis of the predestination of development be looked for, in which one egg distinguishes itself from the others; and this search will be possible with the microscope—how far, no one can say, but its goal is nothing less than a true morphology of heredity.[38]

These views seem to have had a strong influence on August Weismann.

AUGUST WEISMANN AND THE "GERM PLASM"

August Weismann (Figure 4.6) acknowledged the importance of previous attempts to develop a theory of heredity, including Darwin. However, he believed that the whole foundation of the theory of Pangenesis should be abandoned. In the Preface of his book where he presented his own theory of heredity, he noted its main difference from Darwin's theory: not all parts of the body contributed to the substance from which a new individual arose, but rather the offspring owed their origin to a "peculiar substance of extremely complicated structure, viz., the 'germ-plasm'". This could never form anew but only multiply and be transmitted across generations. Based

FIGURE 4.6 August Weismann (1834–1914). Wellcome Collection. Attribution 4.0 International (CC BY 4.0).

on this, Weismann concluded: "My theory might therefore well be denominated *'blasto-genesis'*—or origin from a germ plasm, in contradistinction to Darwin's theory of *'pangenesis'*—or origin from all parts of the body".[39] The contrast between "pangenesis" and "blastogenesis" was an insightful way to describe the main differences between these theories, which are illustrated in Figure 4.7. For Weismann, inherited characters came only from the germline (à la Galton) and not from the whole body, due to specialized sexual cells that were capable of reproducing in the offspring all the characteristics of the parental body. Among those who developed a theory of

(a) Charles Darwin's Pangenesis Theory

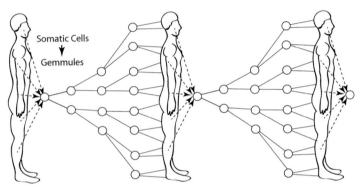

Environment ➤ Modified Gemmules ➤ Inheritance of Acquired Traits

(b) August Weismann's Germplasm Theory

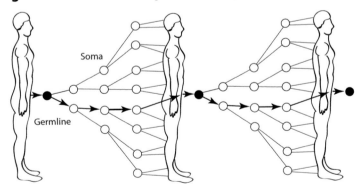

Environment ➤ No Inheritance of Acquired Traits *sensu* Lamarck

FIGURE 4.7 Representation of Darwin's Pangenesis (a) and Weismann's germ plasm theory (b). Whereas according to Darwin's theory germ cells emerged from gemmules from the whole body (hence "Pan-", which means "everywhere", in Pangenesis), Weissman argued that germ cells emerged only from the germline. He thus suggested calling his theory "Blastogenesis". Reproduced with permission from Niklas and Kutschera (2015).

heredity during the latter half of the 19th century, Weismann was the first to seriously consider, and integrate in his theory, the relevant findings from the study of cells. He did this while changing his ideas over time, and adapting them to the new findings.

The new knowledge and understanding of cytology presented in the previous section probably had a major influence on Weismann in developing his theory of heredity. In an essay published in 1883, titled "On Heredity", he described in some detail the main premise of his theory: "We have an obvious means by which the inheritance of all transmitted peculiarities takes place, in *the continuity of the substance of the germ-cells, or germ-plasm*". Weismann believed that the "germ-plasm" had existed continuously since the origin of life, that it was distinct from the main substance of the body, "the somatoplasm", and that changes in the latter could be only due to corresponding changes in the former.[40] In 1885, Weismann proposed the continuity of the germ plasm as the foundation of a theory of heredity. In later years he acknowledged that the fundamental idea had already been suggested by Galton in 1872, "without however being fully appreciated at the time or having any influence on the course of science".[41]

For Weismann, the key question about heredity was how it was possible for a single cell to reproduce all the features of a parent, "with all the faithfulness of a portrait?".[42] He suggested that heredity was due to the transfer across generations of a molecular substance, the "germ-plasm", which possessed a highly complex structure and the power to develop into a complex organism. A key assumption was that part of the germ plasm was not used for the production of the body of the offspring but was rather reserved for the formation of the germ-cells of the following generation. This had an important implication: the continuity of germ cells of successive generations, which could explain similar outcomes.[43] Weismann rejected the inheritance of acquired characters and believed that an organism could not acquire anything unless it already possessed the predisposition to acquire it. In this sense, acquired characters were just local variations and predisposed reactions under certain external influences.[44]

Weismann's theory was presented in a complete form in 1892, in his book titled *The Germ Plasm*. He suggested that all hereditary phenomena depended on minute vital units, the "biophors". He explained that he chose this term, as none of the names used in previous attempts to develop a theory of heredity ever since Spencer was appropriate. Biophors possessed "the power of growth and of multiplication by fission".[45] He nevertheless considered the "pangenes" of de Vries as corresponding almost exactly to biophors, as "*the bearers of the cell-qualities*, the only difference being that the latter were "constituents of higher units of the hereditary substance".[46] The key idea was that all living matter consisted of biophors, and therefore the differences observed in the various tissues were due to the differences among the biophors that existed in the respective cells. This in turn entailed, according to Weismann, that there existed an unlimited number of biophors. This was possible because the biophors were not molecules but groups of molecules; therefore, biophors could vary because of the variations in the set and proportions of molecules of which they consisted.[47]

After stating that biophors constituted all protoplasm, Weismann made a key distinction between two types of it: the "idioplasm" within the nucleus and the "morphoplasm" within the rest of the cell, also assuming that the former had a much more complex structure than the latter. The idioplasm could be involved in development, as it had the ability to regularly undergo changes during growth. When an animal's ovum divided, the initial two embryonic cells were formed, which subsequently generated cells with distinct compositions throughout embryonic development. The variation among these cells, in turn, depended upon modifications in the nuclear material. This eventually raised the question: what was the nature of these changes in the idioplasm, and how were they brought about?[48]

According to Weismann, the structures that constituted the specific character of a cell could arise only by the migration of material particles of the nucleus into the cell body (an idea that he attributed to de Vries). Thus, the development of an undifferentiated embryonic cell to a differentiated one was determined by the presence of the respective biophors. *"Hence the nuclear matter must be in a sense a storehouse for the various kinds of biophors which enter into the cell-body and are destined to transform it"*.[49] Thus, differentiation to specific cell types depended on the different types of biophors present in the cell. In higher organisms, biophors were arranged in groups, in a certain order, all together forming a second order of vital units, the determinants, which controlled the histological character of every cell in a multicellular organism. The different determinants of the germ plasm constituted another unit that Weismann called id, the aggregates of which he called idants.[50]

The process by which the "hereditary substances" of the germ cells of the parents were united into one in the offspring was called "Amphimixis". This process, actually the mixing of two germ plasms, was a source of variation as it might give rise to many different combinations of determinants, which were subjected to the action of natural selection. If this process was repeated in every generation, a doubling of these hereditary substances would take place each time. However, this was not happening due to the reducing division that divided the number of ids, present in the cell, to half. This division took place before the two germ cells were united and secured the constancy of the number of the ids in every generation.[51]

Weismann believed that after fertilization, one part of the germ plasm remained inactive and unchanged and would later produce the germ cells that give rise to the next generation. He also proposed that the characteristics of the offspring were determined by the maternal and paternal ids contained in the germ cells that united in fertilization. As a result, the combination of parental and ancestral characters was predetermined and could not be modified by subsequent influences. In this sense, reversion to remote ancestors was the consequence of the fact that the idants and the ids were not formed anew in the parental germ plasm but were derived from the grandparents. In addition, a "reducing division" produced diversified combinations of ids in the individual germ cells of the parents. Thus, the number of ids of any ancestor, contained in every germ cell, depended entirely on the way this division had occurred.[52]

Weismann's suggestion that the only hereditary variations were those that resulted from the modification of certain determinants of the germ plasm did not imply that external influences could not produce hereditary variations. On the contrary, Weismann believed that they always gave rise to such variations after they had modified some determinants of the germ plasm. New variation was produced by external influence and was established by means of natural selection, which did not act on a single character but on combinations of characters produced by amphimixis.[53] In a sense, Weismann accepted the inheritance of acquired germ plasm variations and rejected the inheritance of the acquired somatic characters.[54] Whereas it may look as if Galton's stirp anticipated Weismann's germ plasm, this is not strictly accurate. Galton took as a fact that the inheritance of acquired characters is rare and unimportant, from which he inferred that inheritance ought to take place from embryo to embryo. Weismann, in contrast, developed a reverse argument. For him, the continuity of the germ plasm was the fact; therefore, the inheritance of acquired characters could not be possible.[55]

The writings of Galton and Weismann that we have considered in this chapter are crucial for understanding what changed during the 34 years between the publication of Mendel's paper in 1866 and its "rediscovery" in 1900: they produced a new context for biological research and a framework of "hard" heredity, in which Mendel's paper was interpreted differently than before.

NOTES

1 Galton (1865), p. 157.
2 Galton (1865), p. 318.
3 Galton (1865), p. 322.
4 Galton (1865), p. 163.
5 Bulmer (2003), p. 104.
6 Galton (1865), p. 319.
7 Galton (1865), p. 322.
8 Galton (1872), pp. 398–399.
9 Galton (1872), p. 400.
10 Bulmer (2003), p. 122.
11 Bulmer (2003), pp. 123–124.
12 Galton (1876), pp. 329–330.
13 Galton (1876), p. 330.
14 Galton (1876), p. 330.
15 Galton (1876), p. 331.
16 Galton (1876), pp. 340–341.
17 Galton (1876), p. 338.
18 Galton (1876), p. 343.
19 Galton (1876), p. 344 (emphasis in the original).
20 Galton (1876), p. 346.
21 Galton (1876), p. 336.
22 Darwin Correspondence Project, "Letter no. 10305," accessed on 15 March 2023 www.darwinproject.ac.uk/letter/?docId=letters/DCP-LETT-10305.xml
23 Darwin Correspondence Project, "Letter no. 10309," accessed on 15 March 2023, www.darwinproject.ac.uk/letter/?docId=letters/DCP-LETT-10309.xml
24 Bulmer (2003), p. 127.
25 Galton (1889), pp. 187–188.
26 Galton (1889), p. 29.
27 Galton (1889), p. 30.
28 Galton (1889), p. 32.
29 Galton (1889), p. 95.
30 Galton (1889), p. 98.

31 Bulmer (2003), pp. 184–186.
32 Galton (1876), p. 334.
33 Galton (1889), p. 187.
34 Galton (1897).
35 Bulmer (2003), ch.8; Pearson (1898).
36 Churchill (2015), pp. 247–253.
37 Harris (1999), pp. 153.158; Paweletz (2001).
38 Flemming (1882), p. 70, quoted in Churchill (2015) p. 260.
39 Weismann (1893), p. xiii.
40 Weismann (1889a), p. 104.
41 Weismann (1904), p. 411.
42 Weismann (1889b). p. 165
43 Weismann (1889b), p. 168.
44 Weismann (1889b), p. 169.
45 Weismann (1893), p. 42.
46 Weismann (1893), p. 42.
47 Weismann (1893), p. 43.
48 Weismann (1893), p. 45.
49 Weismann (1893), p. 48 (emphasis in the original).
50 Weismann (1893), pp. 57,62–63.
51 Weismann (1893), pp. 235–238.
52 Weismann (1893), pp. 335–338.
53 Weismann (1893), pp. 413–418.
54 Winther (2001).
55 Bulmer (2003), p. 132.

5 The Reification of the "Lonely Genius"

WAS NÄGELI REALLY TO BLAME FOR MENDEL'S NEGLECT?

As we saw in the Preface, many prominent geneticists of the first half of the 20th century blamed Nägeli for Mendel's work being unappreciated before 1900. This attitude has continued to recent times. For instance, in her popular account of Mendel's life, titled *The Monk in the Garden*, Robin Marantz Henig wrote: "Because of misdirection, either accidental or deliberate, from the great Nägeli, Mendel lost faith in his own results".[1] In another popular account of Mendel's work, Siddhartha Mukherjee wrote: "His [Mendel's] letters to Nägeli grew increasingly despondent. Nägeli replied occasionally, but the letters were infrequent and patronizing. He could hardly be bothered with the progressively lunatic ramblings of a self-taught monk in Brno".[2] Nägeli was perhaps the only one of his contemporary scientists who came to know about Mendel's work. Despite their long correspondence from 1866 to 1873, the story goes, that Nägeli failed to realize the importance of Mendel's discovery, and this was a major cause for Mendel's work falling into oblivion until 1900. Following Nägeli's advice, Mendel worked on *Hieracium* (hawkweed, a genus of the sunflower family) from 1866 to 1871 but failed to find an agreement between his results and conclusions with this plant and those with *Pisum*.[3] This has been perceived as a bad suggestion by Nägeli that misled Mendel and sent him in the wrong direction research-wise.

Carl Wilhelm von Nägeli (Figure 5.1) was a Swiss botanist. He finished the Zurich gymnasium and then entered the University of Zurich with the intention of studying medicine, where he is said to have been influenced by Lorenz Oken, who taught the *Naturphilosophie*. He subsequently moved to Geneva where he studied botany under Alphonse De Candolle. He finished his doctoral dissertation in 1840 and then spent some time during 1841 at Berlin studying philosophy. In 1842, Nägeli went to Jena and worked with Schleiden, who turned Nägeli's attention to studies with microscopes. He thus focused on the cellular basis of plant growth, and he discovered that dividing cells existed at the apex of shoots and roots of plants. After spending some time in England, he returned to Zurich where he became a lecturer at the veterinary school and soon Professor Extraordinarius. In 1852, he became a professor at Freiburg, and in 1855 he became a professor of general botany at the Polytechnic of Zurich. It should be noted that by that time he had experienced a temporary failure of his eyesight attributed to extensive microscopic work. Finally, in 1857 he became a professor of botany at Munich, a post he held until his death in 1891.[4]

In 1884, Nägeli published his ideas on heredity in his book *A Mechanico-Physiological Theory of Organic Evolution*. He suggested that certain organic compounds formed incipient crystals or micellae. These existed in the plasma and under certain conditions could produce new life, through a process of spontaneous

DOI: 10.1201/9781032449067-6

generation. He also suggested that the larger part of the unarranged, soft, and homogenous primordial plasma of the organism became soma-plasm, with unarranged and easily movable micellae, whereas the smaller part was converted into idioplasm. This was the hereditary substance of the organism, which underwent a "phylogenetic perfecting process", and its configuration eventually became more complex and assumed an adaptation corresponding to external conditions. The distinction between the soma-plasm and the idioplasm was important; however, Nägeli thought that the idioplasm consisted of long strands that went from cell to cell rather than residing inside their nuclei. He also thought that as a particular group of micellae of the idioplasm produced a particular phenomenon in the organism, the former should be designated as the determinant (Anlage) of the latter. Thus, the organism should contain at least as many determinants in its idioplasm as there were different phenomena in it, and if new phenomena appeared in it, new groups of micellae must have previously been introduced into the idioplasm, or the orientation and arrangement of clusters already present must have been changed. Each individual grew up from a cell in which a small quantity of idioplasm was contained. During the subsequent cell divisions, the idioplasm divided into a number of parts, equivalent to that of cells, while it increased in quantity. The idioplasm contained all the determinants that the particular individual had inherited in the germ cell.[5]

FIGURE 5.1   Carl Wilhelm Nägeli (1817–1891). Public Domain, Wikimedia Commons.

According to Nägeli, although the idioplasm preserved its configuration during the development of the organism, new variations occurred because external stimuli might cause local changes, which also reproduced themselves throughout the entire idioplasm. As a result, the germ cells that were produced at any point inherited the effects of those local changes. But it was only changes that involved modifications of the idioplasm that would appear in successive generations; characters acquired otherwise were not inherited. Variation could also be the outcome of sexual reproduction. In the idioplasm of a cell arising from the cross between unlike individuals, the micellar rows of the individual determinants had sometimes an intermediate constitution, thus producing in the offspring characters intermediate between those of the parents. Other times, the micellar rows derived from the father and the mother, respectively, could lie next to one another unchanged in distinct groupings in the idioplasm of the offspring and might reproduce, or not, their respective characters in the organism. In short, variation was produced by external stimuli, which caused modifications of the idioplasm, and by sexual reproduction, which might produce a rearrangement of the determinants in the idioplasm.[6] Finally, the idioplasm contained determinants that always developed, others that developed alternatively, and some that developed only under favorable circumstances. The viable determinants were always, after a certain time, forced into the latent condition. These determinants might be revived and thus rendered capable of development in a descendant. This could explain reversion, the reappearance of ancestral characters.[7]

So, here was a naturalist who was clearly interested in the study of heredity. Indeed, it seems that Nägeli became interested in plant hybridization because he thought that its study could shed light on reproduction and how the transmission of characters from parents to offspring took place. Mendel probably became aware of Nägeli's work on hybridization through Unger, who had mentioned Nägeli in his publications. Mendel sent a reprint of his 1866 paper to Nägeli on December 31 of the same year, and they had a long correspondence until 1873. This consists of ten letters written by Mendel, which were made public in 1905 by Carl Correns (Figure 5.2), a student of Nägeli and one of Mendel's "rediscoverers". In a letter to plant breeder Herbert Fuller Roberts, Correns wrote on January 30, 1925:

> Besides through Focke's book,[8] I had been made cognizant of Mendel's investigations through my teacher Nägeli. And I believe also to remember that he told me of Mendel, but certainly only of the *Hieracium* investigations, in which alone he was permanently interested. . . . The above-cited references to Mendel, and indeed also the recollection of the verbal mention of Mendel, prompted me to ask Nägeli's family for possible letters received. His scientific correspondence was, however, not to be found at the time. It first came to light through an accident in 1904. The letters of Mendel were sent to me by the family, and were published by me; the remaining scientific correspondence the family then destroyed.[9]

Mendel's letters to Nägeli are an interesting reading.[10] Already in the first letter of December 31, 1866, it becomes clear that one of Mendel's motivations for contacting Nägeli was that he had already decided to study *Hieracium*: "*Hieracium, Cirsium*, and *Geum* I have selected for further experiments". Mendel described various difficulties he had encountered with these plants and concluded his first letter by writing:

FIGURE 5.2   Carl Correns (1864–1933). Courtesy of the American Philosophical Society.

> I am afraid that in the course of my experiments, especially with *Hieracium*, I shall encounter many difficulties, and therefore I am turning confidently to your honor with the request that you not deny me your esteemed interest when I need your advice.[11]

So it does not seem to be the case, as the criticisms of Nägeli often have it, that it was he who suggested to Mendel to work with *Hieracium*. Rather, it seems that Mendel had already decided to work with this genus and turned to Nägeli, who was an expert on it for advice.

In his second letter of April 18, 1867, Mendel expressed concerns about his own work:

> With respect to the essay which your honor had the kindness to accept, I think I should add the following information: the experiments which are discussed were conducted from 1856 to 1863. I knew that the results I obtained were not easily compatible with our contemporary scientific knowledge, and that under the circumstances publication of such an isolated experiment was doubly dangerous; dangerous for the experimenter and for the cause he represented. Thus I made every effort to verify, with other plants, the results obtained with *Pisum*. A number of hybridizations undertaken in 1863 and 1864 convinced me of the difficulty of finding plants suitable for an extended series of experiments, and that under unfavorable circumstances years might elapse without my obtaining the desired information.[12]

So here we have Mendel himself acknowledging that his results with *Pisum* could not be confirmed with other plants, something he had already (partially and perhaps hesitantly) acknowledged in his paper where he reported the results with *Phaseolus*. From his letter, we may infer that Nägeli was quite skeptical about Mendel's results, given that Mendel wrote: "I am not surprised to hear your honor speak of my experiments with mistrustful caution; I would not do otherwise in a similar case".[13] But what could Nägeli's reaction have been?

Even though most of Nägeli's letters to Mendel have not been found, part of Nägeli's initial response is available in a letter written on February 25, 1867. Nägeli wrote:

> It seems to me that your experiments with *Pisum* are far from being completed and are indeed only just beginning. A shortcoming of all the more recent experimenters is they have shown so much less perseverance than Kölreuter and Gärtner. I am pleased to note that you are not making this mistake and that you are following in the footsteps of your celebrated predecessors. However, you should try to excel them, and in my view this will only be possible—and this is the only way to make advances in the theory of hybridization—if exhaustive experiments are conducted upon a single object in every possible direction. . . . Your plan to include other plants in your experiments is excellent, and I am convinced that you will obtain quite different results (with regard to the inherited characters) from these other varieties. I should think it particularly valuable if you were able to undertake hybrid fertilizations in *Hieracium*.[14]

Three important points are made by Nägeli in this passage. The first one is that Mendel is on the right path by conducting detailed experiments with many different crosses, as Kölreuter and Gärtner had done in the past. Given their work, the results that Mendel had presented in his 1866 paper could have only been the beginning of such an investigation, if he wished to conduct "exhaustive experiments". The second point is that Nägeli is explicit about the requirement to examine other plants and certainly not only *Hieracium* despite Nägeli's interest in that, if Mendel's results could be obtained for other plants as well—something that Nägeli himself clearly expressed doubts about. Whatever findings Mendel had obtained for *Pisum* would not have been of much broader value if they were restricted to that species only. And what is this all for? According to Nägeli, and this is the third point in the previous quotation, the goal was "to make advances in the theory of hybridization", not a theory of heredity. Nägeli considered Mendel to be working on hybridization because this is what he was actually doing as we saw in Chapters 1 and 2.

In a letter written on November 6, 1867, Mendel reported some of his preliminary observations working with *Hieracium*. There he wrote that he lacked other species, which were necessary for his experiments, going on to explain why this was the case.[15] So we see again not only that it was not Nägeli who suggested to Mendel to focus on *Hieracium* but also that Mendel himself intended to study other plants besides that. In a letter written on February 9, 1868, Mendel wrote to Nägeli that after having acquired some experience in the artificial fertilization of *Hieracia* over the past two years, he intended to conduct some systematic experiments with this genus. To this end, Mendel asked Nägeli where he could buy seeds of the species he did not have. Nägeli seems to have sent Mendel all the seeds he needed, as Mendel thanked him about this in a subsequent letter of May 4, 1868. In that same letter, Mendel also informed Nägeli that he had been elected Abbot of his monastery, which he hoped would provide him with more time to devote to his hybridization experiments.[16]

Mendel continued to report findings and difficulties from his experiments in his letters to Nägeli. In a letter of July 3, 1870 (the eighth out of a total of ten), Mendel reported detailed results from his experiments with *Hieracium*. After reporting that the second and the third generations of various hybrids flowered but not varied as he would have expected, Mendel wrote:

> On this occasion, I can not resist remarking how striking it is that the hybrids of *Hieracium* show a behavior exactly opposite to those of *Pisum*. Evidently, we are here dealing only with individual phenomena, that are the manifestation of a higher, more fundamental law.[17]

It was much later that it was realized that *Hieracium* plants can reproduce through apomixis or apogamy (asexual reproduction in plants, called "parthenogenesis" in animals) and not only by fertilization. Therefore, hybrids did not always form.

It seems therefore that Nägeli did not mislead or misguide Mendel. Rather what happened was that Mendel studied a special case, peas, and more specifically an artificial system that he had himself created between 1854 and 1856. This system was useful in providing new insights for further research, but insufficient for Mendel's stated goal in his paper, to find a law valid for the development of hybrids. Regarding Nägeli's possible inability to appreciate the novelty of Mendel's work, with respect to mathematical analysis, this also seems not to be accurate. Nägeli was one of the most competent persons to understand Mendel's approach. Like Unger, who most likely had an enormous influence on Mendel's work, Nägeli was a botanist positively inclined toward the use of mathematics in scientific research. These "botanical mathematics" were different from the mathematical sciences of that time, and the equations used were mere formalisms that indicated Unger and Nägeli's search for correlations. But mathematics was known to botanists of the time and was used by them. Unger could have thus been the main inspiration for Mendel's mathematical approach in the early 1850s, whereas Nägeli was the next person to turn to in the mid-1860s.[18]

An additional reason for Nägeli's caution against Mendel's results could have been previous cases of mathematical reasoning in botany that were deemed insufficient. For instance, in the 1830s, the German botanist Carl Schimper conceived a mathematical formula for predicting where leaves would sprout on a growing plant stem. It seems that the law that Mendel was seeking to establish was vulnerable to the same kinds of criticisms that Schimper's "law" received in the late 1850s and early 1860s: that mathematics did not mirror the internal workings of the plant. Predicting leaf positions could be correct by coincidence or for the wrong reasons, and it could not be accepted by botanists investigating the growth and development of plants at the cellular level. Eventually, it was Nägeli's microscopy studies of leaf formation that provided the most important evidence against Schimper's "law", which was thus reduced to a mere empirical rule of thumb. It is possible that Nägeli had considered Mendel's work in the same light as Schimper's: predicting hybrid ratios successfully was not sufficient for Nägeli to accept a law guiding the development of hybrids.[19]

This is, as we saw earlier, why he asked Mendel to do more crosses and with more species.

Finally, Nägeli did actually refer to Mendel's 1866 paper in 1885 and to Mendel's work on *Hieracium* in 1891. Most importantly, the Brno Natural Science Society had sent more than 100 copies of the journal that included Mendel's paper to several scientific academies around the world, both in Europe and in the United States. Besides Nägeli's, there were 13 more references to Mendel's paper before 1900, and some of them were in books that were widely read by naturalists. The Royal Society's Catalogue of Scientific Papers (1864–1873) volume 8, published in 1879, cites three of Mendel's papers. Hermann Hoffmann referred to Mendel's paper in a little book he published in 1869. Most notably, Mendel's work was also cited in Wilhelm Focke's book, which was a basic reference for hybridization studies.[20] The real story is certainly more complicated, and interesting, than what the stereotypical narrative about the "lonely-genius being ahead of his time" suggests.

One last point worth adding is Nägeli's possible positive influence on Weismann. According to historian Frederick Churchill, also Weismann's main biographer, Nägeli's work clearly shaped Weismann's ideas, and understanding Nägeli's work therefore helps one gain a considerable understanding of Weismann's work. Nägeli believed that the phenomena of heredity depended on some material substance, that a mechanism of heredity could be found through chemistry, and that a theory of heredity should make sense physiologically. Most importantly, Nägeli had explicitly distinguished the transmission of characters from the development of characters and identified each of these two processes with unique activities of the molecular substance in cells. Of course, Nägeli got many things wrong. But, as Churchill put it, by envisioning an idioplasm independent of the rest of the organism, he fulfilled the fundamental requirements of a mechanical theory of both intergenerational and intragenerational continuity.[21]

MENDEL'S "REDISCOVERY" IN 1900

In 1900 Carl Correns (Figure 5.2), Hugo de Vries (Figure 5.3), and Erich von Tschermak (Figure 5.4) published their results on plant hybridization that were in agreement with those obtained by Mendel.[22] As we saw in Chapter 2, de Vries was one of the naturalists who tried to develop a theory of heredity during the latter part of the 19th century. This theory was first proposed in 1889, in his book *Intracellular Pangenesis*.[23] Even though Weismann's germ plasm theory was published in its complete form in 1892 after the publication of de Vries' *Intracellular Pangenesis*, his original ideas had become known earlier and had influenced de Vries. As the title of the book indicates, de Vries accepted Darwin's hypothesis that individual hereditary characters depended on individual material particles in the living substance of cells. However, he rejected the idea of the transportation of gemmules through the entire body. He called those particles "pangens" and suggested that all living protoplasm was built up of these pangens, of which there was one for every hereditary character. According to de Vries, every kind of pangen was represented in the nucleus in an inactive form; only those that were related to nuclear division became active. However, pangens could multiply in either the active or the inactive state. The

FIGURE 5.3   Hugo de Vries (1848–1935). Courtesy of the American Philosophical Society.

remaining protoplasm, which de Vries called cytoplasm, contained those that were going to become functional in the specific cell type. He assumed that the same particles were alternatively activated and inactivated, depending on the circumstances. Occasionally, inactive pangens were transported from the nucleus to certain organs of the cytoplasm where they were activated in order to perform certain functions. In every cell of higher organisms, certain groups of pangens dominated and impressed their character on it.[24]

De Vries made a distinction between different cell lines: the primary germ tracks, the secondary germ tracks, and the somatic tracks. He considered as primary germ tracks those sequences of generations of cells that normally led from the fertilized ovum to the new germ cells, in both animals and plants. The secondary germ tracks existed only in plants and consisted of cells that could develop into new individuals and meristematic cells that could give rise to buds. Finally, the somatic tracks gave rise to the cells of the grown individual, except from the germ cells.[25] De Vries applied Weismann's distinction between somatic tracks and germ tracks to likewise reject the inheritance of acquired characters.[26] De Vries identified two kinds of variability, which corresponded to those of Darwin. "Fluctuating variability" was based on the varying ratios of the individual kinds of pangens, which could be changed "by their multiplication and under the influence of external circumstances, but most quickly by breeding selection". On the other hand, "species-forming variability" resulted from the fact that when divided the pangens produced two new ones that were usually similar to the original one, but that could rarely be dissimilar. Both forms would then multiply, and the new one would "tend to exercise its influence on the visible characters of the organism".[27] De Vries undertook a large number of experiments that finally convinced him that there was a strong distinction between

fluctuating variations having no evolutionary significance and what he called mutations by discontinuous leap.[28]

It seems that de Vries had found the 3:1 ratio by 1895 in crosses of the species
Papaver somniferum (commonly known as opium poppy); however, he did not mention it in the International Conference on Hybridisation that took place in London
in July 1899. Yet, in March 1900, de Vries wrote two papers in which he discussed
the 3:1 ratio. Did his views change in the meantime? Historian Robert Olby has
argued that the notion of paired characters cannot be found in the writings by de
Vries before 1900. But this changed, likely due to what occurred between July 1899
and March 1900. First, de Vries received Mendel's paper from his friend Professor
Beijerinck, who was aware that de Vries was working on hybrids and who thought
that he might find Mendel's 1866 paper useful. Second, Correns, who as we saw was
Nägeli's student and thus already aware of the work of Mendel, published a paper
in December 1899 in which he mentioned Mendel's work on peas without however
citing his paper. It is therefore likely that between July 1899 and March 1900, de
Vries read carefully Mendel's paper, which made him interpret his own hybridization
experiments in a new light.[29] Interestingly, de Vries' account of his "discovery" of
Mendel's results, in a letter to Roberts on December 18, 1924, did not mention any of
these events. Rather he wrote:

> After finishing most of these experiments, I happened to read L.H. Bailey's 'Plant
> Breeding' of 1895. In the list of literature of this book, I found the first mention of
> Mendel's now celebrated paper, and accordingly looked it up and studied it.[30]

So de Vries himself stated that he knew of Mendel's paper since 1895.

In March 1900, de Vries sent two papers to the French journal *Comptes Rendus
de l' Académie des Sciences* and the German journal *Berichte der deutschen bota-
nischen Gesellschaft*. However, in the *Comptes Rendus* paper, which had the title
"Sur la loi de disjonction des hybrides" ("On the law of the segregation of hybrides"),
de Vries did not mention Mendel. In contrast, he did so in the *Berichte* paper, which
was longer and more extensive compared to the two-page-long *Comptes Rendus*
paper. In the *Berichte* paper, de Vries wrote the following in a footnote about
Mendel's paper: "This important treatise is so seldom cited, that I myself for the first
time came to know about it *after I had closed the majority of my experiments, and
had derived therefrom the principles contributed in the text*".[31] With this comment, it
seems that de Vries wanted to clarify that it was not Mendel's paper that had guided
his research; notwithstanding the importance of Mendel's paper, de Vries considered
himself to have arrived at the same results independently. But when exactly de Vries
read Mendel's paper is not clear. According to historian Robert Olby, he had probably
come across Mendel's paper various times, so his account in the letter to Roberts
may be confused. However, it also seems that de Vries did not feel that he had any
intellectual debt to Mendel and wanted to suppress any such idea.[32]

Whatever the motivation of de Vries was, the lack of any reference to Mendel
in the *Comptes Rendus* paper initiated an immediate reaction by Correns. As he
described in his own account of what happened, in a letter he wrote to Roberts on
January 23, 1925:

On the morning of the 21st of April, 1900, I received a separate "Sur la loi de disjunction des hybrids" of de Vries, and by the evening of the 22nd of April, my contribution, "G. Mendels Regel über das Verhalten der Nachkommenschaft der Bastarde" ("G. Mendel's law on the behavior of progeny of hybrids") was ready.[33]

Correns submitted his paper to the *Berichte* too, and it was published in May. But why did Correns rush to publish his paper, submitting it one day after reading that of de Vries, and why did he mention Mendel's name in its title? According to sociologist Augustine Brannigan, Correns was probably long aware of Mendel's results, even though as he admitted in his letter to Roberts he could not recall when exactly he had read Mendel's paper. As the *Comptes Rendus* paper by de Vries did not mention Mendel, Correns might have thought that de Vries was claiming Mendel's results as his own. Correns thus rushed to write and submit his own paper to *Berichte*, and perhaps he described the results as "Mendel's Law" in order to undermine de Vries' priority and to insist on Mendel's priority over both de Vries and himself. In other words, Mendel was brought back to the scene to resolve a potential priority dispute.[34]

What exactly Correns wrote at the beginning of his *Berichte* paper to achieve this is quite interesting:

> The latest publication of Hugo De Vries, . . . which I came into possession of yesterday, through the generosity of the author, prompts me to the following contribution: . . . I also, in my hybridization experiments with races of maize and peas, had arrived at the same results as De Vries, who experimented with races of very different sorts of plants, among them also with maize races. When I had found the orderly behavior, and the explanation therefor—to which I shall immediately return—it happened in my case, as it manifestly now does with De Vries, that I held it all as being something new. *I then, however, was obliged to convince myself that the Abbot Gregor Mendel in Brünn in the 60's, through long years of and very extended experiments with peas, not only had come to the same result as De Vries and I, but that he had also the same explanation, so as far as it was possible in 1866.*[35]

Correns was explicit in the paper about his views on priority. Even though he was long aware of Mendel's work, he did not worry much about being credited with its "rediscovery": "I have then not held it to be necessary to assure myself the priority for this 'post-discovery', through a preliminary contribution, but decided to continue the experiments still further".[36] After analyzing Mendel's paper from a 1900 perspective, writing about "sexual nuclei" that correspond to characters, Correns concluded: "I call this the Mendelian Rule; it includes also De Vries' 'Loi de disjunction'. Everything further may be derived from it".[37] In short, Correns rushed to let the scientific community know that Mendel got it right long before anyone else, including de Vries and himself, and therefore should be credited with the "discovery" of this "law". It was Mendel's law.

What about the third "rediscoverer" Erich von Tschermak (Figure 5.4)? As already mentioned, he was the grandson of Eduard Fenzl, one of Mendel's professors in Vienna. He was also the youngest and less known among Mendel's "rediscoverers". Here is his own account of the "rediscovery" in a letter to Roberts written on January 7, 1925, describing his work in 1899:

In counting out the seed characters, the ever-recurring number relationship of 3:1 could naturally not escape me, any more than the number relation of 1:1 on back-crossing of green-seeded peas with hybrid pollen of the F_1 generation. . . . From this standpoint, I designated the regularities found by Mendel and myself, as those of the regularly varying values of the characters of inheritance.[38]

Tschermak went on to say that in autumn 1899 he read through Focke's renowned book on hybridization, published in 1881. In the section of that book dedicated to peas, he noticed a reference to Mendel's paper. Tschermak found it and read it, and he was surprised to realize that Mendel had actually reported the very same ratios in 1866. However, Tschermak thought of himself as being the only one at the time who had discovered Mendel's work. In January 1900, he submitted a paper to his university in the form of a dissertation. But when at the beginning of April 1900 he received from de Vries, whom he had visited in 1898, his *Comptes Rendus* paper, he realized that de Vries was also aware of Mendel's work. Referring to a sentence in de Vries' paper about dominant and recessive characters, Tschermak wrote:

I read this sentence with the greatest interest, but also, frankly stated, with consternation, for it was now quite clear to me that De Vries must also know the work of Mendel, although it was not cited in this paper. For me it was naturally, as a beginning docent, not indifferent that my work should be anticipated, wherefore I immediately sought from the rectorate the permission to let my already censored work be taken out and printed.[39]

FIGURE 5.4 Erich von Tschermak-Seysenegg (1871–1962) soon after the "rediscovery" of Mendel's paper. Archiv der Österreichischen Akademie der Wissenschaften, Nachlass Erich von Tschermak-Seysenegg, P-0183-A.

Tschermak thus rushed to publish his own work. After reading de Vries' *Berichte* paper, he also became aware of Correns' paper and was surprised anew that someone else was also aware of Mendel's work. He thus prepared a paper, which he considered an abstract of his work, titled "Über künstsliche Kreuzung bei Pisum sativum" ("Concerning artificial crossing in *Pisum sativum*").[40] He submitted it to the *Berichte* as well, and it was received on June 2. Tschermak considered Mendel's paper so important that he requested its inclusion in the series Ostwalds *Klassiker der exakten Wissenschaften* (Ostwald's *Classic Texts in the Exact Sciences*), where it indeed appeared as Volume 121 in 1901.[41]

Perhaps the "rediscovery" of Mendel's work is less heroic and romantic than what we would have liked it to be. At least three botanists had found similar results with one another. When they realized that Mendel had arrived at these results long before them, they reacted differently. de Vries did not mention Mendel at all in his first paper to appear in print. Perhaps he did so because his own paper was only two pages long. Or he did not think that Mendel's work was really that important. The latter seems a more likely explanation for his omission than an intention to overlook Mendel's work in order to claim priority.[42] Yet this omission was enough for Correns to react immediately and insist on Mendel's importance and priority, thus bringing him back to the scene.

BATESON FOUNDS "MENDELIAN GENETICS"

William Bateson (Figure 5.5) was the next one to argue for the importance of Mendel's work. He had been influenced by Galton and William Keith Brooks (under whom he had previously studied for some time) and considered discontinuous

FIGURE 5.5 William Bateson (1861–1926). John Innes Archives courtesy of the John Innes Foundation.

variation as having enormous importance, certainly being more important than continuous variation that Darwin had favored. In 1899 he had met de Vries in person, during the International Conference on Hybridization of the Horticultural Society in England. This was the beginning of a friendship, in part due to their common interest in hybridization and their advocating of discontinuous variation.[43] In a talk given at that conference, Bateson had already proposed the kind of research that would be necessary in order to understand heredity and evolution:

> What we first require is to know what happens when a variety is crossed with its nearest allies. If the result is to have a scientific value, it is almost absolutely necessary that the offspring of such crossing should then be examined statistically. It must be recorded how many of the offspring resembled each parent and how many showed characters intermediate between those of the parents. If the parents differ in several characters, the offspring must be examined statistically, and marshalled, as it is called, in respect of each of those characters separately.[44]

While traveling from Cambridge to London in May 1900, in order to give a talk at the meeting of the Royal Horticultural Society, Bateson read the *Berichte* paper by de Vries. As soon as he finished, he modified his talk, initially titled "Problems of Heredity as a Subject for Horticultural Investigation", in order to include some of Mendel's results. Finding out about Mendel's work had an immediate and long-lasting impression on him.[45] Bateson thus immediately became an "apostle of Mendel". He was also the person who introduced Mendel's work to the English-speaking world, as he commissioned the first English translation of Mendel's paper by Charles Druery, which was published in the journal of the Royal Horticultural Society in 1901.[46] Bateson also wrote an introductory note for this translation:

> The conclusion which stands out as the chief result of Mendel's admirable experiments is of course the proof that in respect of certain pairs of differentiating characters the germ-cells of a hybrid, or cross-bred, are pure, being carriers and transmitters of either the one character or the other, not both. That he succeeded in demonstrating this law for the simple cases with which he worked it is scarcely possible to doubt. In so far as Mendel's law applies, therefore, the conclusion is forced upon us that a living organism is a complex of characters, of which some, at least, are dissociable and are capable of being replaced by others. We thus reach the conception of unit-characters, which may be rearranged in the formation of the reproductive cells. It is hardly too much to say that the experiments which led to this advance in knowledge are worthy to rank with those that laid the foundation of the Atomic laws of Chemistry.[47]

Thus Bateson affirmed the importance of Mendel's work. It must be noted that it was Bateson who introduced the concept of "unit-characters", which as we see in subsequent chapters came to be the predominant concept of early 20th-century genetics. This is an additional reason, to that mentioned in Chapter 1 that what became known as Mendelian genetics is actually Batesonian genetics. What people still today read in an anachronistic manner in Mendel's paper was in reality Bateson's interpretation and integration of that paper into the science of his day.

Despite his insistence on the importance of Mendel's work, Bateson cautiously added: "To what extent Mendel's conclusions will be found to apply to other characters, and to other plants and animals, further experiment alone can show".[48] He was confident however that the work of Mendel would form the basis of something new and important:

> Nevertheless, however much it may be found possible to limit or to extend the principle discovered by Mendel, there can be no doubt that we have in his work not only a model for future experiments of the same kind, but also a solid foundation from which the problem of Heredity may be attacked in the future.[49]

Despite his confidence, Bateson suggested: "The whole paper abounds with matters for comment and criticism, which could only be profitable if undertaken at some length".[50] Such a criticism would soon be published by Weldon and frustrate Bateson (see Chapter 6).

It is worth noting here that, despite his crediting of Mendel for the discovery of his "law", Correns had written toward the end of his *Berichte* paper, published in May 1900:

> At present, however, this law is applicable only to a certain number of cases, i.e., those where one member of a pair of traits dominates, and probably only to hybrids between varieties. It seems impossible that all parts of traits of all hybrids should behave according to this law. Some hybrids of peas bear this out.[51]

On October 30, 1901, de Vries wrote to Bateson:

> I prayed you last time, please don't stop at Mendel. I am now writing the second part of my book which treats of crossing, and it becomes more and more clear to me that Mendelism is an exception to the general rule of crossing. It is in no way *the* rule! It seems to hold only in derivative cases, such as real variety characters.[52]

So, what we see here are two of the "rediscoverers" suggesting caution about Mendel's "law". Correns made it clear in the very first paper he wrote about Mendel's results that they were not widely applicable. De Vries went even further by suggesting to Bateson that what Mendel had found was applicable in very few cases. But Bateson was thrilled about Mendel's paper.

In what followed, thanks to Bateson, Mendel's paper became the foundational document of the new science of genetics. But why? When Mendel's paper was read in a new context after 1900, it was given a new meaning that was immediately put into practice. Historians Staffan Müller-Wille and Giuditta Parolini have provided an explanation for the quick acceptance of Mendel's paper by examining readers' annotations when they read it. A striking finding from the examination of a sample of annotated copies of Mendel's paper was that annotations are found in particular parts of the paper. The sections toward the end of the paper, where a theoretical explanation for the empirical results of the previous sections is proposed, usually lack annotations. In contrast, the first sections of the paper that explain the material used and its experimental arrangement and then report the empirical findings of the crosses have attracted the most attention, with the sections on "Selection of experimental plants" and "Arrangement and order of the trials" being the most heavily annotated ones. Moreover, Mendel's statement that purebred plants possessing constantly differing traits must be used for hybridization experiments was almost universally highlighted. Other practical significant details, such as Mendel's warning that the green coloration of the albumen sometimes is faint, making the peas appear yellow, were also often annotated. For Müller-Wille and Parolini such annotations, along with the lack thereof in the more theoretical sections, indicated that for many readers Mendel's paper was first and foremost a practical "how-to" guide for conducting experiments. It seems therefore that Mendel's paper provided the grounds for developing an experimental system that combined pure-breeding and artificial cross-fertilization with a mathematical notation to record, analyze, and visualize experimental results.[53]

In the 1900s, the work of Mendel guided the development of the new science of "genetics," a term coined by Bateson. He mentioned the term "genetics" for the first time in a letter of April 18, 1905, to a friend, noting that for "a professorship relating to Heredity and Variation. . . . No single word in common use quite gives this meaning. Such a word is badly wanted and if it were desirable to coin one, 'Genetics' might do".[54] The term appeared the next year in a book review that Bateson wrote. He also proposed the term in 1906, during his inaugural address to the Third Conference on Hybridization and Plant Breeding of the Royal Horticultural Society. The term was adopted for the published proceedings the next year, describing the event as the Third International Conference on Genetics.[55]

Thus, the stage for a new science was set—but not by Mendel himself. It was the reading of Mendel's paper in a new context in 1900 that paved the way for genetics. Therefore, it is important to clearly distinguish between the pre-1900 period, when Mendel's paper had no impact whatsoever on any of the theories of heredity developed during that period and the post-1900 period when Mendel's paper had a major impact.

NOTES

1 Henig, (2000), p. 161.
2 Mukherjee, (2016), p. 55.
3 Olby (1985), p. 103.
4 Vines (1880); Scott (1891).
5 Nägeli (1898/1884), pp. 2–13.
6 Nägeli (1898/1884), pp. 16–21.
7 Nägeli (1898/1884), pp. 24–25.
8 This was a basic reference book for hybridization studies.
9 Roberts (1929), p. 338.
10 Mendel's letters to Nägeli form the body of a paper Correns published in 1905, which was reprinted in a book by him in 1924. They were published in English in the journal *Genetics* in 1950, but here I draw on the translation in Stern and Sherwood (1966).
11 Stern and Sherwood (1966), pp. 58–59.
12 Stern and Sherwood (1966), p. 60.
13 Stern and Sherwood (1966), p. 61.

14 Quoted in Stubbe (1972), pp. 158–159.
15 Stern and Sherwood (1966), p. 74.
16 Stern and Sherwood (1966), pp. 78–79.
17 Stern and Sherwood (1966), p. 90.
18 Dröscher (2015).
19 Gliboff (1999).
20 Olby (1985), pp. 216–220.
21 Churchill (2015), Chapter 10.
22 English translations were published in the journal *Genetics* in 1950 (De Vries 1950; Correns 1950; Tschermak 1950).
23 de Vries (1910).
24 de Vries (1910), 193–207.
25 de Vries (1910), 93–103.
26 de Vries (1910), 210.
27 de Vries (1910), 214; see also Stamhuis et al. (1999).
28 Lenay (2000).
29 Olby (1985), pp. 113–116.
30 Quoted in Roberts (1929), p. 323.
31 Roberts (1929), pp. 324, 328.
32 Olby (1985), p. 116.
33 Quoted in Roberts (1929), p. 337.
34 Brannigan (1981), pp. 94–95; see also Brannigan (1979) for the first account.
35 Quoted in Roberts (1929), p. 339.
36 Quoted in Roberts (1929), p. 339.
37 Quoted in Roberts (1929), p. 342.
38 Quoted in Roberts (1929), p. 345.
39 Quoted in Roberts (1929), p. 346.
40 Tschermak (1950).
41 Roberts (1929), p. 346; see also Tschermak (1951).
42 Brannigan (1981), p. 95; Olby (1985), p. 116.
43 Provine (2001), p. 66.
44 Bateson (1899).
45 Cock and Fordsyke (2022), p. 170.
46 Olby (2000).
47 Mendel (1901), p. 1.
48 Mendel (1901), p. 1.
49 Mendel (1901), p. 2.
50 Mendel (1901), p. 2.
51 Stern and Sherwood (1966), p. 131.
52 Quoted in Provine (2001), p. 68.
53 Müller-Wille and Parolini (2020).
54 Quoted in Dunn (1991/ 1965), p. 69.
55 Dunn (1991/ 1965), pp. 68–69; Olby (2000).

Part II

Social Mendelism

6 Mendel's Great Defender

WELDON CRITICIZES "MENDELISM"

We saw in the previous chapter that both Correns and de Vries suggested caution about Mendel's "law". However, the first detailed critique of Mendel's work came from Walter Frank Raphael Weldon (Figure 6.1). Weldon and Bateson were initially friends, as they both had studied during the early 1880s in Cambridge. Weldon was elected a fellow of the Royal Society of London in 1890 and Bateson in 1894, so during the 1890s they often encountered each other in the respective meetings. A discussion of Weldon's work in an 1895 meeting resulted in the deterioration of their increasingly frosty relationship. The situation became even worse when Galton invited Bateson to join the Royal Society committee on the statistical study of evolution, despite Weldon's objections. Eventually, Weldon and others left the committee in early 1900.[1] In the meantime, Weldon and his colleague and friend in University College London Karl Pearson (more on whom in Chapter 8) founded the journal *Biometrika*, and it was in its pages that in January 1902 Weldon's paper titled "Mendel's Laws of Alternative Inheritance in Peas" appeared ("alternative" meaning "non-blending").[2]

Weldon initially stated Mendel's two laws, as he perceived them, which he called the laws of "Dominance" and "Segregation":

> The first general result obtained by Mendel may be stated as follows: *If peas of two races be crossed, the hybrid offspring will exhibit only the dominant characters of the parents; and it will exhibit these without (or almost without) alteration, the recessive characters being altogether absent, or present in so slight a degree that they escape notice.*[3]
>
> This may be called the Law of Dominance, and it at once explains the terms "dominant" and "recessive".
>
> The second result is that: *If the hybrids of the first generation, produced by crossing two races of peas which differ in certain characters, be allowed to fertilise themselves, all possible combinations of the ancestral race-characters will appear in the second generation with equal frequency, and these combinations will obey the Law of Dominance, so that characters intermediate between those of the ancestral races will not occur.*[4]
>
> From its consequences, this may be called the Law of Segregation.

In commenting on Mendel's results, Weldon noted that among 8,023 seeds of the second generation, 6,022 were yellow and 2,001 green, with no seeds of intermediate color occurring. He thus remarked that "The ratio between either of these numbers and the number of seeds observed is an excellent approximation to that required by Mendel's law of segregation".[5] Weldon then proceeded to consider what one should have found if one considered deviations, due to chance, of the observed results from those expected based on theory. After making the relevant calculations,

FIGURE 6.1   Walter Frank Raphael Weldon (1860–1906) in the garden. *Source:* Pearson 5/1/5. UCL Special Collections, UCL Archives, London.

he concluded that Mendel's results in terms of the agreement between the expected and the observed frequencies were overall very good. One could then conclude that Mendel's own results supported his theory. But Weldon concluded, by applying Pearson's statistical methods, that "the odds against a result as good as this or better are 20 to 1".[6] In other words, Mendel's results were not all that probable.

But that was not all. Even if Mendel's results supported his theory, how widely applicable the laws deduced from those results were? Weldon wrote:

> It is almost a matter of common knowledge that they do not hold for all characters, even in Peas, and Mendel does not suggest that they do. At the same time I see no escape from the conclusion that they do not hold universally for the characters of Peas which Mendel so carefully describes.

He noted that in suggesting so, he had no intention of belittling the importance of Mendel's achievement but only to point to fruitful areas of further research.[7]

Weldon considered the "Law of Dominance" first. He noted that the evidence about dominance in the hybrids of the first generation mostly concerned the color of cotyledons and seed coats, as well as the shape of the seeds. But he also cautioned that one should carefully consider what the statement that a character is dominant really meant. Many varieties of peas were very variable, both in color and in shape. For instance, a variety with "round smooth" seeds did not always produce identical seeds. At the same time, others were not entirely smooth, and in some cases all seeds had differences.

> So that both the category "round and smooth" and the category "wrinkled and irregular" include a considerable range of varieties. At the same time, the categories are

undoubtedly often discontinuous, the most wrinkled seed of such a race as *Express* or *Victoria* being so much smoother and more rounded than the most regular seed of the typically "wrinkled" races, that no one who knows both races would hesitate for a moment in deciding which race a given seed resembled.[8]

Weldon thus argued that contrary to Mendel's assumptions about the existence of "either/or" characters, that is, character pairs of forms that are clearly distinct from each other, there was so much variation that the designated categories, say, "round and smooth" or "wrinkled and irregular", were not internally homogeneous. Instead, they showed considerable variation, despite often seemingly being discontinuous. The dominance of "smooth" over "wrinkled" seeds meant that when an individual that belonged to the category "smooth-seed" was crossed with an individual that belonged to the category "wrinkled-seed", all the offspring would belong to the category "smooth-seed", but this did not mean that those offspring would be exactly the same with one another.

Weldon went on to question the universality of dominance of yellow over green as a pattern, by discussing studies with peas that showed exceptions to Mendel's law. Then he noted that if the dominance of the yellowness over the greenness of the cotyledons in the hybrids was universal, those who first worked on peas, such as Knight (mentioned in Chapter 2), would have certainly noticed it. Yet, despite the numerous crosses they had performed, none of them had made such a claim.[9] Weldon then turned to the shape of the seeds and argued that the available evidence made the idea of dominance of round over wrinkled peas also difficult to sustain. Even more difficult was the case to be made for the color of the seed coats. After presenting evidence from the work of Tschermak, Correns, and others, Weldon concluded:

> that dominance of any of the characters mentioned is not an invariable attribute of the character, but that a cross between pairs of parents, such that the different members of each pair differ to the same extent in cotyledon colour or in similar characters, may in different cases lead to widely different results.[10]

He then proceeded to explain why this was the case. His view was that whatever characters an organism exhibited did not only depend on those of the parental generation alone but also on those of its more remote ancestors. This is an implicit reference to Galton's Law of Ancestral Heredity, according to which an individual had inherited features from all its ancestors, with each contribution being lower the more remote an ancestor was (see Chapter 4). But, Weldon noted, "Mendel does not take the effect of difference of ancestry into account, but considers that any yellow-seeded Pea, crossed with any green-seeded Pea, will behave in a certain definite way, whatever the ancestry of the green and yellow peas may have been".[11] According to Weldon, Mendel had assumed that any two yellow-seeded peas would be the same, but this was not necessarily so as there might be differences due to their different ancestries. Ancestry mattered, and this was why according to Weldon the particular patterns of dominance reported by Mendel were not universal. Weldon concluded:

> These facts show first that Mendel's law of dominance conspicuously fails for crosses between certain races, while it appears to hold for others; and secondly that the intensity of a character in one generation of a race is no trustworthy measure of its dominance

in hybrids. The obvious suggestion is that the behaviour of an individual when crossed depends largely upon the characters of its ancestors. When it is remembered that Peas are normally self-fertilised, and that more than one named variety may be selected out of the seeds of a single hybrid pod, it is seen to be probable that Mendel worked in every case with a very definite combination of ancestral characters, and had no proper basis for generalisation about yellow and green peas of any ancestry.[12]

He also emphasized some relevant observations that Correns had made.

Weldon then suggested that if Mendel's conclusions were valid at least for peas, the characters of the various hybrids should fall into one or the other of a few definitive categories, without any intermediate forms. However, based on his own observations, he had concluded that neither that was the case. In contrast, he had found that many intermediate forms existed. Based on a color classification on the top row of a plate included at the end of his article (Figure 6.2), Weldon presented his results in the form of a table (Table 6.1): among seeds of various types, many had fairly uniform color, but an important number in each case were "obviously pie-bald", that is, consisted of different colors.

Most interestingly, the existence of this variation is something that Mendel himself had also acknowledged in his paper:

> However, part of the traits just listed do not allow for a certain and sharp separation, since the difference often rests on a "more or less" that is often difficult to determine. Such traits were unsuitable for the individual trials; these had to confine themselves

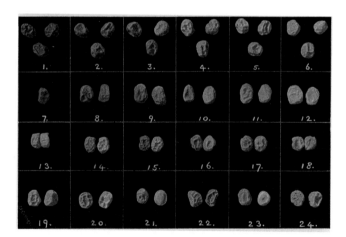

FIGURE 6.2 Weldon's reference plate for classifying seeds according to color. "Figs. 1–6. Seeds of the hybrid Pea Telephone, the seed-coats removed, arranged so as to form a colour scale. Figs. 7–12. Seeds of the hybrid Pea Stratagem, with the coat removed, forming a colour scale. Figs. 13–18. Each figure shows a pair of cotyledons belonging to one seed of the Pea Telephone. The series shows the various degrees of difference in colour between the two cotyledons of the same seed. Figs. 19–24. Peas in their seed-coats, showing the way in which the cotyledon colour is masked. 19–20, Telephone; 21, Telegraph; 22, Stratagem; 23, Pride of the Market; 24, Early Morn". Reproduced with permission from Weldon (1902), p. 255.

TABLE 6.1
Frequency of Color in Pea Varieties Variation

Race	Color 1	Color 2	Color 3	Color 4	Color 5	Color 6	Piebald	Total
Telegraph	354	95	47	10	4	2	64	576
Pride of the Market	447	76	19	2	2	1	55	602
Stratagem	200	367	154	16	5	0	40	602
Telephone (Carter)	191	289	195	59	38	38	133	943
Telephone (Sutton)	13	83	112	32	15	13	43	311
Telephone (Vilmorin)	29	69	69	23	3	2	5	200
Duke of Albany	26	70	121	53	11	20	27	328
Daisy	78	175	27	7	0	0	17	304
Early Morn	267	239	81	2	1	1	9	600

Source: Based on TABLE IV in Weldon (1902).

> to those characteristics that emerge clearly and decisively in the plants. In the end, success had to show whether they [i.e., the selected characters] all follow a concordant behaviour in hybrid union, and whether from this a judgement becomes possible about those traits too that have a subordinate typical significance.[13]

Thus even Mendel himself had acknowledged that he had somehow "cherry-picked" pea varieties that fulfilled his criteria for clear-cut characters. This is a perfectly legitimate choice for a model system but limits significantly the generalizability of the results.

We therefore see how, already in 1902, Weldon showed that Mendel's "laws" of inheritance might not even be valid for peas. Weldon's studies of varieties of pea hybrids led him to conclude that there was a continuum of colors from greenish yellow to yellowish green, as well as a continuum of shapes from smooth to wrinkled in gradually increasing degrees. In Weldon's view, Mendel had not shown how heredity in general works but had rather studied a special case. What Mendel had in fact done was to create a model system; as a model for the study of heredity it could be great, but it was not generalizable. Recall from Chapter 1 that Mendel devoted two whole years, between 1854 and 1856, in order to obtain purebred plants for his experiments. But it seems that in doing so, Mendel actually eliminated all the natural variation that existed in peas.

Weldon's criticism was legitimate, but Bateson did not think so.

BATESON FIGHTS BACK—FIERCELY

Bateson did not leave Weldon's critique unchallenged. In 1902, he published a short book that included a response to Weldon, along with a modified version of Druery's translation of Mendel's paper. That short book was titled *Mendel's Principles of Heredity: A Defence*,[14] in which Bateson overall presented Mendel's work as providing the solutions for various problems relevant to heredity. In the Preface of the book, Bateson wrote:

But every gospel must be preached to all alike. It will be heard by the Scribes, by the Pharisees, by Demetrius the Silversmith, and the rest. Not lightly do men let their occupation go; small, then, would be our wonder, did we find the established prophet unconvinced. Yet, is it from misgiving that Mendel had the truth, or merely from indifference, that no naturalist of repute, save Professor Weldon, has risen against him?[15]

In a nutshell: Mendel found the truth about how heredity works, and only Weldon had reacted against him—either due to misgiving or due to indifference. This could only harm science, according to Bateson. Had the criticism come from a junior scholar, it might have been overlooked, but coming from Weldon, it would certainly influence younger scholars and might thus turn them away from the related field of research. This is why Bateson decided to defend Mendel against Weldon, in order to point out where the latter "has gone wrong, what he has misunderstood, what omitted, what introduced in error".[16]

Bateson went on to argue that the kind of work that Mendel did, based on experimentation and statistical analysis, was what would make biology an exact science. But for some reason Weldon was fast to cancel this "first positive achievement of the precise method",[17] Bateson remarked. He essentially blamed Weldon that he stood against the advancement of biology by turning against the best-established methods, and one could figure out that Weldon was wrong if one had the patience to compare his arguments against the original studies. But this was not all. Bateson continued:

With sorrow I find such an article sent out to the world by a Journal bearing, in any association, the revered name of Francis Galton, or under the high sponsorship of Karl Pearson. I yield to no one in admiration of the genius of these men. Never can we sufficiently regret that those great intellects were not trained in the profession of the naturalist.[18]

Not only Weldon was wrong, but those who supported him, like Pearson and Galton, could not have realized this because they were not trained naturalists like Bateson. None of them knew what they were talking about, Bateson implied.

The next chapter was a modified version of a paper titled "The Problems of Heredity and Their Solution" that Bateson published in 1900 right after finding out about Mendel's work. Bateson argued that whereas Galton's Law of Ancestral Heredity was a crucial first step toward describing the respective phenomena, it could not account for all phenomena. This happened, for instance, in cases of what he described as "discontinuous variation": "cases in which actual intermediates between the parent forms are not usually produced on crossing".[19] Bateson emphasized that Mendel studied discontinuous characters, provided an outline of his findings, and then noted that Correns and Tschermak had confirmed those same findings. Then he proceeded with the key contribution that Mendel made, where we can trace the origins of the anachronistic reading of Mendel's work that we find in the stereotypical narrative of Mendelian genetics:

In the simplest case, suppose a gamete from an individual presenting any character in intensity A unite in fertilisation with another from an individual presenting the same character in intensity a. For brevity's sake we may call the parent individuals A and a, and the resulting zygote Aa. What will the structure of Aa be in regard to the character we are considering?

Up to Mendel no one proposed to answer this question in any other way than by reference to the intensity of the character in the progenitors, and primarily in the parents, A and a, in whose bodies the gametes had been developed.[20]

So, here is the problem: Bateson, based on his own experience and understanding as a naturalist, as well as on the knowledge of his time, found in Mendel's paper what he considered to be an answer to a problem that was puzzling naturalists. But instead of stating that it was his own reading of Mendel's paper in a new context that made him see in it an experimental approach that could be fruitful for research, he stated that this was what Mendel himself had achieved. As we saw in Chapter 1, this is far from accurate.

Moving on, Bateson argued that the Law of Ancestral Heredity could work well for cases of blending inheritance but not for cases of non-blending, or alternative, inheritance. As he put it, "*it does not directly attempt to give any account of the distribution of heritage among the gametes of any one individual.* Mendel's conception differs fundamentally from that involved in the Law of Ancestral Heredity".[21] Bateson showed why this was the case while introducing some key terms of Mendelian genetics, which are sometimes mistakenly attributed to Mendel himself:

Consequently if *Aa*'s breed together, the new *A* gametes may meet each other in fertilisation, forming a zygote *AA*, namely, the pure *A* variety again; similarly two *a* gametes may meet and form *aa*, or the pure a variety again. But if an *A* gamete meets an a it will once more form *Aa*, with its special character. This *Aa* is the hybrid, or "mule" form, or as I have elsewhere called it, the *heterozygote*, as distinguished from *AA* or *aa* the *homozygotes*.[22]

And

Each such character, which is capable of being dissociated or replaced by its contrary, must henceforth be conceived of as a distinct *unit-character*; and as we know that the several unit-characters are of such a nature that any one of them is capable of independently displacing or being displaced by one or more alternative characters taken singly, we may recognize this fact by naming such unit-characters *allelomorphs*. So far, we know very little of any allelomorphs existing otherwise than as pairs of contraries, but this is probably merely due to experimental limitations and the rudimentary state of our knowledge.[23]

The fundamentals were now in place. Depending on what allelomorphs (a term that was later replaced by the shorter term "allele", introduced by George Shull in 1927) an individual had, it would be described as a homozygote if it carried the same one twice, or as a heterozygote if it carried two different ones. Having introduced such key terms, Bateson went on to elaborate the principles of Mendelian genetics as they were established thereafter, and as they are still taught in schools today.

The chapters that included the translations of Mendel's papers (on *Pisum* and *Hieracium*) were followed by a 105-page-long chapter titled "A Defence of Mendel's

Principle of Heredity". There Bateson argued against Weldon's critique. However, here I want to focus on two points only: (1) Bateson's response to Weldon's criticism that Mendel's work was a special, artificially created case that was not generalizable, and (2) the various explicit and implicit accusations against Weldon outside the technical aspects of plant hybridization and heredity. The former I consider important because it is the reason that in the Conclusions, I argue against teaching the stereotypical of Mendelian genetics in schools; and the second in order to show that Bateson's response was not an exchange between peers but rather a clash of personalities.

Bateson began by expressing his admiration for the work of Galton that had led to the Law of Ancestral Heredity. Yet, in his view that law could not account for some phenomena observed, which could rather be explained by Mendel's Law. Whether or not the latter could be expanded to also account for those cases accounted for by Galton's law, it was for Bateson a question that might be answered in the future—and for him there was no reason why this would not be the case. This notwithstanding, the unquestionable fact for Bateson was that every case that followed Mendel's law was forever to lay outside the realm of the Law of Ancestral Heredity. Given this, Bateson noted: "It is not perhaps to a devoted partisan of the Law of Ancestral Heredity that we should look for the most appreciative exposition of Mendel".[24] Bateson thus suggested that Weldon was biased because what he was supporting (Galton's law) had been challenged, and therefore he could not have been expected to give an unbiased assessment of Mendel's work. Indeed, Bateson continued, Weldon did not mention in his critique a large, and perhaps most significant, part of Mendel's experiments; neither did he mention Mendel's conception of the hybrid as a distinct entity with its own characters, which was for Bateson the most novel idea in the whole paper. Instead, Bateson continued, Weldon selected dispersed facts and statements that suggested that dominance was not widely applicable, as well as that Mendel's method was fundamentally flawed as it did not take ancestry into account. Bateson commented, "To find a parallel for such treatment of a great theme in biology we must go back to those writings of the orthodox which followed the appearance of the 'Origin of Species' ".[25] Bateson went on to address the specific criticisms of Weldon in different sections.

The first section was titled "I. The Mendelian Principle of Purity of Germ-Cells and the Laws of Heredity Based on Ancestry". Bateson began with a historical account of the development of the Law of Ancestral Heredity, providing statements as evidence that the law was not universally applicable. He described his own understanding of the law like this: "*It is an essential part of the Galton-Pearson Law of Ancestral Heredity that in calculating the probable structure of each descendant the structure of each several ancestor must be brought to account*".[26] Interestingly, in a footnote on the same page Bateson remarked that he did not have a precise understanding of that law, because of the mathematical form in which it was expressed, but had "every confidence that the arguments are good and the conclusion sound".[27] Bateson then explained why that law could not account for the cases in which Mendel's principles applied. The crossbreeding of pure varieties did not have to "diminish the purity of their germ-cells or consequently the purity of their offspring". Consequently, when the hybrids bred with one another, or with the parental pure forms, the offspring that would emerge were "stated to be as pure as if they had had no cross in their pedigree, and henceforth their offspring will be no more likely to depart from the *A* type or the *B* type respectively, than those of any other originally pure specimens of these types". Therefore, the ancestry of the individuals did not really matter, because they were "pure" and so the Law of Ancestral Heredity was not applicable.[28] It should be noted though that for either/or characters, specific versions of the Law of Ancestral Heredity—for homozygous dominant or recessive individuals—could be shown to follow Mendel's laws.[29]

In the next section, titled "II. Mendel and the Critic's Version of Him: The 'Law of Dominance' ", Bateson addressed Weldon's criticisms on what the latter described as the Law of Dominance. He essentially suggested that Weldon was fighting a strawman, as Mendel himself had not stated any such law. "There is as yet no universal law here perceived or declared",[30] Bateson noted. Weldon, Bateson continued, put emphasis on the supposed Law of Dominance, while overlooking the purity of the germ cells, which was the key fact because it made ancestry less important than what Weldon had suggested. So, according to Bateson, Weldon invented a problem that did not exist: attributed to Mendel a law that he had not proposed himself and then proceeded to show that that law was not supported by the available evidence. "We are asked to believe that Professor Weldon has thus discovered 'a fundamental mistake' vitiating all that work, the importance of which, he elsewhere tells us, he has 'no wish to belittle' ".[31]

The following section was "III. The Facts in Regard to Dominance of Characters in Peas", and it begins like this:

> Professor Weldon refers to no experiments of his own and presumably has made none. Had he done so he would have learnt many things about dominance in peas, whether of the yellow cotyledon-colour or of the round form, that might have pointed him to caution.[32]

Weldon had mostly studied animals rather than plants and certainly did not have Bateson's botanical experience and expertise. Bateson thus emphasized some key distinctions. The first one was between the color of the seed cotyledon that was a character derived from the embryo and the color of the seed coat that was a maternally derived character. Furthermore, there were varieties with opaque coats that masked the color of the cotyledon and transparent coats that did not mask the color of the cotyledon.[33] By the way, the seeds shown in Plate I in Weldon's 1902 critique (Figure 6.2) all had their seed coats removed. Bateson also considered seed shape, pointing out that sometimes it can be influenced by external conditions. He then devoted considerable space to discuss the cases studied by Mendel and discussed by Weldon. Toward the end, Bateson concluded:

> The soundness of Mendel's work and conclusions would be just as complete if dominance be found to fail often instead of rarely. For it is perfectly certain that varieties can be chosen in such a way that the dominance of one character over its antagonist is so regular a phenomenon that it can be used in the way Mendel indicates. He chose varieties, in fact, in which a known character *was* regularly dominant and it is because he did so that he made his discovery. When Professor Weldon speaks of the existence of fluctuation and diversity in regard to dominance as proof of a "grave discrepancy"

between Mendel's facts and those of other observers, he merely indicates the point at which his own misconceptions began.[34]

What about the non-binarity of characters suggested by Weldon? Bateson accepted Weldon's observations with *Telephone* but considered it "a good example of an extreme case of mixture of both colours and shapes". He also added that "regular dimorphism in respect of shape is not so common as dimorphism in respect of colour".[35] For Bateson, *Telephone* was an "impure green", "very irregular in colour, having many intermediates shading to pure yellow and many piebalds". This was also the variety from which alone Weldon had made his color scale, Bateson added, implying that it was not sufficient to challenge Mendel's results.[36] *Telephone* was an exception to the rule, according to Bateson, "an impure and irregular green".[37] Bateson came back to this point again and again: "the colour-heredity of *Telephone* is abnormal makes it fairly clear that there is here something of a really exceptional character. What the real nature of the exception is, and how far it is to be taken as contradicting the 'law of dominance', is quite another matter".[38] This is really impressive if one considers what Bateson famously stated in 1908:

> Treasure your exceptions! When there are none, the work gets so dull that no one cares to carry it further. Keep them always uncovered and in sight. Exceptions are like the rough brickwork of a growing building which tells that there is more to come and shows where the next construction is to be.[39]

Weldon's untimely death in 1906 ended the debate. But Bateson would soon have to deal with more kinds of exceptions.

TREASURING EXCEPTIONS

According to Bateson's reinterpretation of Mendel's paper, when two heterozygotes for one character were crossed (Aa × Aa), the phenotypic ratio in the offspring would be 3:1, with three out of four offspring exhibiting the dominant character and one out of four the recessive character (see also Figure 1.2). Accordingly, when two heterozygotes for two characters were crossed (AaBb × AaBb), the phenotypic ratio in the offspring would be 9:3:3:1. In particular, 9 out of 16 offspring would exhibit the two dominant characters, 3 out of 16 one dominant and one recessive character, 3 out of 16 the other dominant and the other recessive characters, and 1 out of 16 the two recessive characters (see also Figure 1.3). However, as early as 1901 Bateson and his close collaborator Edith Saunders (Figure 6.3) reported deviations from the Mendelian ratios, including cases where particular hereditary characters in plants did not segregate independently from each other.

One such case was the plant *Matthiola*. Two pairs of characters that seemed to be related to pairs of alleles were "hoary" (hairy leaf surface) vs. "glabrous" (smooth leaf surface), and "green" vs. "brown" seed color. There were cases that followed simple Mendelian principles but also others that did not. For instance, "hoary" was generally found to be dominant over "glabrous". When "hoary" (Type 1) and "glabrous" plants were crossed, about 1,000 offspring were produced that were all

FIGURE 6.3 Edith Rebecca Saunders (1865–1945) in her garden allotment. Reproduced with permission from Richmond (2006).

completely "hoary". Interestingly even when "half-hoary" plants (Type 2—plants having only the underside of the leave hoary, with the upper side being "glabrous") were used, the offspring were completely "hoary", similar to the "hoary" offspring emerging from a self-fertilized "hoary" parent. However, there were exceptions to this rule. When two "half-hoary" plants were crossed with "glabrous" ones, they produced different results: the crosses with one "glabrous" plant gave the usual "hoary" offspring, whereas the crosses with another "glabrous" plant gave 72 plants that were all "glabrous". Bateson and Saunders could not explain this result and assumed that it might be due to repeated self-fertilization, admitting that the available evidence was not sufficient to establish this. With respect to seed color, "green" generally appeared to be dominant over "brown". However, Bateson and Saunders remarked that the intensity of the green color could become diminished over crossing.[40] These were interesting exceptions to dominance that Bateson considered as a rule.

In one experiment, the phenotypic ratio in the offspring was 1244 "hoary" to 407 "glabrous"; this is a 3.05:1 ratio that was legitimately described by Bateson and Saunders as "a very close approximation to the Mendelian ratio 3:1". However, they also reported that they had found a "complete correlation between seed colour and leaf-character, green seeds giving rise to hoary, brown and intermediate to glabrous plants". In another experiment, they reported that " green seeds again produced hoary plants, brown seeds, with four exceptions, glabrous plants". The total proportion of plants was 1,429 "hoary" vs. 373 "glabrous", which they described as a 3.8:1 ratio. Such deviations were reported in other experiments. Whereas in one cross "glabrous" × "hoary" a ratio of 117 "hoary" vs. 39 "glabrous" was obtained, which

FIGURE 6.4   Reginald Crundall Punnett (1875–1967). John Innes Archives courtesy of the John Innes Foundation.

Bateson and Saunders described as "an exact Mendelian result", in other crosses there were significant deviations from the expected results. Another cross "glabrous" × "hoary" gave a ratio of 98 "hoary" vs. 18 "glabrous", which is a 5.4:1 ratio; whereas yet another cross "glabrous" × "hoary" gave a ratio of 175 "hoary" vs. 39 "glabrous", that is, an almost 4.5:1 one.[41] For these two cases, Bateson and Saunders concluded that there was a deficiency of offspring exhibiting the recessive phenotype.[42]

There were more exceptions to consider and report along the way. In a report published in 1908, with the involvement of Reginald Punnett (Figure 6.4), another close collaborator of Bateson, a new phenomenon was introduced: epistasis. Generally speaking, this is the phenomenon in which the expression of one gene is modified (for instance, inhibited or suppressed) by the expression of other genes. Bateson and Punnett conducted experiments with poultry, studying the inheritance of comb: the fleshy, red outgrowth on the top of the head in poultry (Figure 6.5). They explained their findings with the "presence and absence" hypothesis. They considered the "rose-comb" as a comb with an additional element "rosiness". They thus suggested that there were two alleles: "presence of a factor for rose" (*R*) and "absence of that factor" (*r*). The rose comb, they explained, was actually a single comb modified by the presence of a "rose" factor, whereas the absence of the factor would result in a single comb. They concluded that the presence of a given factor was dominant to the absence of that factor, and they suggested that this rule applied to all cases of Mendelian heredity studied that far. They also added:

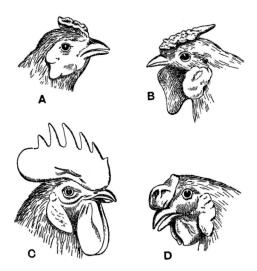

FIGURE 6.5   Fowls' combs: (A) "pea"; (B) "rose"; (C) "single"; (D) "walnut". Reproduced from Punnett (1911), p. 30, fig. 4. Public Domain.

As the acceptance of the "Presence and Absence" hypothesis seems to demand some general expression for such inter-relation between factors belonging to distinct allelomorphic pairs, we propose the terms *epistatic* and *hypostatic*. For example, the combless, the single-combed, and the rose-combed conditions may, in the light of our present knowledge, be regarded as forming a cumulative series, and we should speak of the factor for single as being dominant to the combless condition but hypostatic to the rose factor; and similarly the rose factor may be referred to as epistatic to the single.[43]

This was their way of integrating the new findings into the already available Mendelian (actually Batesonian) scheme. The presence of a factor influenced the effect of another factor that belonged to a different pair of alleles. Let us consider a concrete example.

There were four kinds of comb in poultry, which Bateson, Saunders, and Punnett called "Single-comb", "Rose-comb", "Pea-comb" and "Walnut-comb" (Figure 6.5). Based on their observations, they suggested that the alleles were R for "Rose-comb" and r for its absence, as well as P for "Pea-comb" and p for its absence.[44] Here is what they had previously observed: when they crossed "Rose-comb" with "Pea-comb" individuals, all their offspring had a "Walnut-comb". This was a peculiar case, with respect to Mendel's laws. Not only neither the "Rose-comb" nor the "Pea-comb" were dominant, but also a different character, "Walnut-comb", appeared. When "Walnut-comb" individuals were crossed with each other, the following ratios were observed in the next generation: 9 "Walnut-comb":3 "Rose-comb":3 "Pea-comb":1 "Single-comb". Based on this and other crosses, Bateson, Saunders and Punnett suggested that:

- "Rose-comb" individuals had the allele R but not the allele P.
- "Pea-comb" individuals had the allele P but not the allele R.

- "Walnut-comb" individuals had both the allele *P* and the allele *R*.
- "Single-comb" individuals had neither the allele *P* nor the allele *R*.[45]

Based on these results, they concluded: "The factor for rose is epistatic to the factor for single and the single factor is hypostatic to rose".[46] But what was the factor for single? It was neither *P* nor *R*, but one that belonged to another pair of alleles that was influenced by the allelic pair *P* and *R*. They further clarified this with an explicit definition:

> The term epistatic is thus applied to denote such a relationship between the factors which are not in the same allelomorphic pair. A factor, then, is epistatic to another, when by its presence it conceals the existence of the other factor, although not allelomorphic to it. The terms dominant and recessive should only be applied to express relationship between factors in the same allelomorphic pair.[47]

Bateson and his colleagues soon observed other ratios, such as 15:1 and 9:7. These new epistatic phenomena did not lead to a reconsideration of Mendelism but rather to its modification in order to accommodate the new findings. The model looked too good to be abandoned.

During this work, there was another innovation in the representation of their results. Table 6.2 shows how Bateson, Saunders, and Punnett represented the aforementioned results of the poultry comb. In today's jargon, the table indicates the genotypes with capital letters and the phenotypes ("Walnut-comb", "Rose-comb", "Pea-comb", and "Single-comb") with small letters (r.p., r., p., and s., respectively). This representation provides a nice overview of the results, where one can also easily see the phenotypic ratio 9 "Walnut-comb":3 "Rose-comb":3 "Pea-comb":1 "Single-comb", which stems from the genotypic ratio 9 R_P_:3 R_ _ _:3 _ _ P_:1 _ _ _ _. This shows why they described the phenomenon just with reference to the presence or absence of "rose" or "pea" in 1906.

The representation in Table 6.2, as well as in Figures 1.2 and 1.3, is legendary in Mendelian genetics and is described as the "Punnett square" because it was invented by Punnett. As it is used widely in the teaching of Mendelian genetics, it is worth considering its history and its use.

THE PUNNETT SQUARE

The Punnett square is a square tabular array used in genetics to represent the genotypes (and based on those also the phenotypes) of the progeny from a cross between two organisms. Punnett wrote a short and widely read book titled *Mendelism*, which was first published in 1905. However, the Punnett square did not appear in the 1905 first edition but in the second one of 1907 (and as we saw in a different form in the 1906 report to the evolution committee). Punnett described a typical case of the inheritance of two characters, each of which in turn depended on a pair of "unit-characters" *Aa* and *Bb*, with *A* being dominant to *a* and *B* being dominant to *b*. Crossing an *AABB* individual with an *aabb* one, would yield only *AaBb* individuals in the next generation (F_1). When these were crossed, Punnett continued, we should expect to

TABLE 6.2

Representation of the Results of a Cross in the 1906 Report to the Evolution Committee by Bateson, Saunders, and Punnett

RP RP (r.p.)	RP R, no P (r.p.)	RP no R, P (r.p.)	RP no R, no P (r.p.)
R, no P RP (r.p.)	R, no P R, no P (r.)	R, no P no R, P (r.p.)	R, no P no R, no P (r.)
no R, P RP (r.p.)	no R, P R, no P (r.p.)	no R, P no R, P (p.)	no R, P no R, no P (p.)
no R, no P RP (r.p.)	no R, no P R, no P (r.)	no R, no P no R, P (p.)	no R, no P no R, no P (s.)

Source: Based on Bateson, Saunders, and Punnett (1906), p. 13.

R and P represent factors, the presence or absence of which is thus indicated, whereas r.p., r., p., and s. indicate what we today describe as phenotypes, in this case, "Walnut-comb", "Rose-comb", "Pea-comb", and "Single-comb", respectively.

have the following offspring in F_2: $AA + 2Aa + aa$ and $BB + 2Bb + bb$. These could be combined in all possible ways. Here is what he did: he conceived initially of three kinds of squares—one with vertical lines for the presence of allele A, one with horizontal lines for the presence of allele B, and one with no lines for a or b, indicating the absence of either A or B. Then he ordered the four possible genotypes of the first cross (AA, Aa, aA, aa) in the four parts of the large square and combined them with the four possible genotypes of the second cross (BB, Bb, bB, bb). This resulted in the square in Figure 6.6a.

Statistician-geneticist Anthony Edwards has argued that the initial Punnett square is essentially a multiplication table. The first members of the pairs in the four boxes of successive rows are AB, Ab, aB, ab, and the second member of the pairs in the four boxes of successive columns are similarly AB, Ab, aB, ab. Whereas Bateson had explicitly acknowledged that this representation was introduced by Punnett, Edwards traced further its origin to a diagram that Galton had sent to Bateson on October 1, 1905, in which Galton had reorganized the three-factor data that Bateson had previously sent to him. It seems nevertheless that Punnett already had a clear sense of the usefulness of the square. Another possible influence was John Venn, who invented the Venn diagrams and who must have interacted with Punnett in Cambridge.[48] Whatever the case, it is interesting that the representation in Figure 6.6a is different from the representation typically used in Mendelian genetics, such as that in Figure 1.3. Indeed, the logic of the construction of the Punnett square evolved from the second edition of *Mendelism* of 1907 to the third edition of 1911, where we find

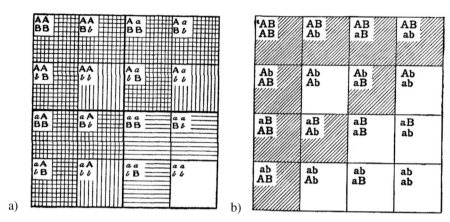

a)   b)

FIGURE 6.6 (a) Punnett square from the second edition of his book *Mendelism* (1907) showing the genotypes of the 9:3:3:1 phenotypic ratio. Reproduced from Punnett (1907), p. 45. Public Domain. (b) Punnett square from the third edition of his book *Mendelism* (1911) showing the genotypes of the 9:3:3:1 phenotypic ratio. Reproduced from Punnett (1911), p. 43. Public Domain.

a more modern representation, shown in Figure 6.6b. In this case, it is clear that the square is organized on the basis of the allelic constitution of the gametes (1st row: AB; 2nd row: Ab; 3rd row: aB; 4th row: ab; 1st column: AB; 2nd column: Ab; 3rd column: aB; 4th column: ab). Perhaps a biological logic (combination of gametes), rather than a mathematical logic (combinatorial mathematics), was more useful eventually. Punnett referred to this new method of constructing the square as the "chessboard" method:

> A convenient and simple method of demonstrating what happens under such circumstances is the method sometimes termed the "chessboard" method. For two series each consisting of four different types of gamete we require a square divided up into 16 parts. The four terms of the gametic series are first written horizontally across the four sets of four squares, so that the series is repeated four times. It is then written vertically four times, care being taken to keep to the same order. In this simple mechanical way all the possible combinations are represented and in their proper proportions.[49]

Because of the differences in the initial and later Punnett squares, as shown in Figures 6.6a and 6.6b, philosopher William Wimsatt has suggested that perhaps Punnett had not fully grasped the principles behind their construction. There was some ambiguity in the interpretation of the initial larger Punnett squares constructed to represent more segregating factors. The interpretation by Punnett, and probably by Bateson, Saunders, and their contemporaries gradually changed from 1907 to 1913, and it is only then that the so-called chessboard method and the new interpretation became widely used. This new way of reading the Punnett square was critical to its emergence as a powerful and adaptable conceptual tool. According to Wimsatt, taking the gametes of each parent as the basic elements to be combined rather than doing the logical assembly of a genotype locus by locus allows one to organize the contributions of each parent to different sides of the array, making both the production of the results and their interpretation easier. Any number of alleles and loci can be represented by listing all the gametic combinations possible at the head of the rows and columns in an array. This could only be done for two alleles based on the initial conceptualization.[50]

The Punnett square seems to have been invented during the experiments on the inheritance of comb in poultry, described in the previous section. It also seems that Punnett presented this experiment in his undergraduate courses and that he encouraged his students to construct the squares themselves. For instance, the lecture notes taken by the biochemist Dorothy Needham, who attended Punnett's undergraduate course in genetics in 1917–1918, include three Punnett squares in the notes for the first lecture, like the one in Table 6.1. Perhaps Punnett thought that students of genetics had to learn by doing and that by constructing the squares themselves they might be led to understand Mendel's laws and the logic of genetics. Indeed, both segregation and independent assortment become clearly visible in a Punnett square, and so its teaching value in this sense is enormous.[51] As an anecdote, I can add from my personal experience that both school teachers and school students like using them. They are fun for students to be engaged with and easy for teachers to assess.

However, despite their usefulness in representing crosses and making complex mathematical data easy to represent, Punnett squares have a fatal flaw when it comes to their interpretation by non-geneticists, including school students: phenotypes are represented as being directly produced by genotypes, thus masking entirely the underlying developmental processes. Depending on the alleles that a particular offspring has, it is possible to thereafter infer its phenotype. So, to take a classic example, from the cross between two *AaBb* individuals that have a [AB] phenotype, it is possible to have offspring with:

- genotype *A_B_* and phenotype [AB]
- genotype *A_bb* and phenotype [Ab]
- genotype *aaB_* and phenotype [aB]
- genotype *aabb* and phenotype [ab]

No further questions are asked. Once a genotype emerges in the Punnett square, the respective phenotype is taken as given. But this is far from accurate.

Even though Bateson's and Punnett's use of the Punnett square conveyed the opposite message, a given genotype does not always result in a given phenotype. The phenotype of an individual may indicate a genotype, but it is developmental processes that actually produce the phenotype. This was already clear to geneticists early in the 20th century, as we see in Chapter 7. Nowadays, we rely on the concepts of penetrance and expressivity to describe the relation between genotype and phenotype. Penetrance is the proportion of individuals with a given genotype who have a typically associated phenotype; in other words, penetrance is the concept indicating how likely a particular phenotype is given a certain genotype. Expressivity is about the qualitative differences in the phenotypes of individuals with the same genotype or in other words the degree to which a particular phenotype appears given a certain genotype. It has been argued though that expressivity and penetrance do not pose a problem for genetics research, as they do not diminish the relevance of Mendelian segregation, dominance or recessiveness (and I would add Punnett squares), in understanding the inheritance of traits. In experimental systems, variation in penetrance and expressivity is quite rare because genetic and environmental variations are controlled. Thus, when exceptional cases are found, they can be attributed to previously unrecognized variability in such factors. When we cannot control either the environmental or the genetic background of the individuals under study, then variation in expressivity and penetrance can occur.[52] Here is then the problem with the representations relies on Punnett squares. They may be useful for experimental systems and the related research done by experts, but they do not accurately represent natural systems and heredity for non-experts. What is missing in the typical representations of Punnett squares are all the assumptions and conditions under which the results presented occurred. For instance, we saw earlier that Weldon showed that Mendel arrived at his results after a two-year procedure during which he practically eliminated all the natural variation in peas. Not distinguishing between experimental and natural outcomes can result in confusion.

Another issue with the Punnett square is that it works well when the factors are segregated and independently assorted. It is only in this way that all the possible combinations of alleles are possible. But by the time Punnett and Bateson were using the Punnett square, it had already been suggested that this may not be always the case. Instead, it had been found that it might even be impossible for some factors to always be inherited independently.

NOTES

1 Radick (2023), pp. 8–10.
2 Weldon (1902).
3 Weldon (1902), p. 229 (emphases in the original).
4 Weldon (1902), p. 229 (emphases in the original).
5 Weldon (1902), p. 230.
6 Weldon (1902), p. 235.
7 Weldon (1902), p. 235.
8 Weldon (1902), p. 236.
9 Weldon (1902), p. 237.
10 Weldon (1902), pp. 240–241.
11 Weldon (1902), p. 241.
12 Weldon (1902), p. 242.
13 Mendel (2020), p. 29.
14 Bateson (1902).
15 Bateson (1902), p. vi.
16 Bateson (1902), p. viii.
17 Bateson (1902), p. xi.
18 Bateson (1902), p. xii.
19 Bateson (1902), p. 7.
20 Bateson (1902), pp. 18–19.
21 Bateson (1902), p. 22 (emphasis in the original).
22 Bateson (1902), p. 23 (italics in the original).
23 Bateson (1902), p. 27 (italics in the original).

24 Bateson (1902), p. 105.
25 Bateson (1902), p. 106.
26 Bateson (1902), p. 110 (emphasis in the original).
27 Bateson (1902), p. 110 (emphasis in the original).
28 Bateson (1902), p. 114.
29 Radick (2023), pp. 221–224.
30 Bateson (1902), p. 118.
31 Bateson (1902), p. 118.
32 Bateson (1902), p. 119.
33 Cock and Fordsyke (2022), p. 189.
34 Bateson (1902), p. 136.
35 Bateson (1902), p. 124.
36 Bateson (1902), p. 146.
37 Bateson (1902), p. 147.
38 Bateson (1902), p. 152.
39 Bateson (1908), p. 19.
40 Bateson and Saunders (1901), p. 35.
41 Bateson and Saunders (1901), p. 69.
42 Bateson and Saunders (1901), p. 77.
43 Bateson, Saunders, Punnett (1908), p. 2.
44 Bateson, Saunders, Punnett (1908), p. 18.
45 Bateson, Saunders, Punnett (1906), pp. 12–13.
46 Bateson, Saunders, Punnett (1908), p. 19.
47 Bateson, Saunders, Punnett (1908), p. 43.
48 Edwards (2012), (2016)
49 Punnett (1911), p. 33.
50 Wimsatt (2012).
51 Green (2019).
52 Botstein (2015), pp. 23–24.

7 Chromosomes, "Factors", and Genes

A CHROMOSOMAL BASIS FOR HEREDITY

By 1900, many cytologists had suggested that particles inside the cells might be related to the emergence of traits. The 3:1 ratio observed by Mendel's "rediscoverers" could then be explained with the assumption that particulate determinants existed in the nucleus, which were segregated and independently assorted. Thus, in 1900 Mendel's paper was seen as bringing together the results of breeding experiments with the facts of cytology.[1] Here is how Tschermak described that in 1951:

> The three rediscoverers were fully aware of the fact that the independent discovery of the laws of heredity in 1900 was far from being the accomplishment it had been in Mendel's time since it was made considerably easier by the work which appeared in the interval, especially the cytological researches of Hertwig and Strasburger.[2]

As we saw in Chapter 4, already during the 1880s, cytologists had assumed that chromosomes were somehow involved in heredity, because of the regularity and the precision with which those were formed and became transmitted at each cell division.

However, it was Theodor Heinrich Boveri (Figure 7.1) who first showed with experimental manipulations, not just observations, in *Ascaris* that the number and the morphology of the chromosomes were maintained during cell division. He described this as the principle of "the Continuity of Chromosomes". Later, with simple procedures, he showed that it was possible to produce urchin blastomeres, the cells of the early embryo, with different and incomplete sets of chromosomes. During his previous study of the centrosomes in *Ascaris*, Boveri had come to the conclusion that each chromosome could attach only to two centrosomes and that this attachment was random and mutually exclusive. Working with sea urchin eggs, he became aware that when they were fertilized in vitro, some zygotes developed immediately into four-cell embryos, called "tetrasters". Boveri had already observed this phenomenon with *Ascaris* and knew that it was caused by two sperm entering the egg. In contrast to a normal zygote where there existed two sets of chromosomes and two centrosomes, the dispermic ones contained three sets of chromosomes (two from the two sperm cells and one from the egg), as well as four centrosomes to which chromosomes could be attached. This provided Boveri with the ability to manipulate the chromosome content of individual sea urchin blastomeres in terms of chromosome type and number. He then let these manipulated embryos develop and the outcomes varied, from cells dying immediately to developing different tissues. His interpretation was that these outcomes were due to different distributions of chromosomes, which in turn suggested that each chromosome carried different hereditary

DOI: 10.1201/9781032449067-9

FIGURE 7.1   Theodor Heinrich Boveri (1862–1915). Wikimedia Commons. Public Domain.

information. Boveri described this as a second principle, that of "the Individuality of Chromosomes". Given the two principles of continuity and individuality of chromosomes, it should be no surprise that Boveri noticed the potential value of cytological work for understanding heredity and made such comments in papers he published between 1902 and 1904.[3]

In 1903, Walter Sutton (Figure 7.2) provided cytological evidence that the rearrangements of chromosomes during cell division could account for Mendel's ratios. He had already mentioned such a connection in a paper published the previous year, where he had written: "I may finally call attention to the probability that the association of paternal and maternal chromosomes in pairs and their subsequent separation during the reducing division as indicated above may constitute the physical basis of the Mendelian law of heredity".[4] There he also expressed his intention to return to this topic, and so he did with a 1903 paper in the beginning of which he stated:

> The general conceptions here advanced were evolved purely from cytological data, before the author had knowledge of the Mendelian principles, and are now presented as the contribution of a cytologist who can make no pretensions to complete familiarity with the results of experimental studies on heredity. As will appear hereafter, they completely satisfy the conditions in typical Mendelian cases, and it seems that many of the known deviations from the Mendelian type may be explained by easily conceivable variations from the normal chromosomic processes.[5]

So, here we have Sutton approaching heredity from a different angle than earlier researchers. Instead of having a breeder who studies how the characters of individuals

FIGURE 7.2 Walter Stanborough Sutton (1877–1916). Seated portrait, circa 1905. *Source*: box 2, folder 1, Walter Stanborough Sutton papers, KUMC MSS I, University of Kansas Medical Center Archives, Kansas City, Kansas.

are transmitted across generations and who speculates about what is going on within cells, we have a cytologist who studies the latter and tries to draw conclusions for the former. Sutton also attributed to Mendel an understanding that we found in the stereotypical narrative of Mendelian genetics, which is likely due to Sutton's reading of Bateson's 1902 book, discussed in Chapter 6. Sutton wrote: "It has long been admitted that we must look to the organization of the germ-cells for the ultimate determination of hereditary phenomena. Mendel fully appreciated this fact and even instituted special experiments to determine the nature of that organization".[6]

Sutton then summarized some conclusions from his study of chromosomes in the reproductive cells of the genus *Brachystola* (grasshoppers). He was concerned about recent observations indicating that the maternal and paternal chromosomes remained distinct in the reproductive cells. According to those, during the "reducing division" (meiosis), all the maternal chromosomes went to one pole of the cells and all the paternal ones to the other. As a result, reproductive cells should contain either the maternal or the paternal chromosomes. But this, Sutton realized, would pose limits to the possible combinations of the hereditary material in the offspring. As he wrote: "If any animal or plant has but two categories of germ cells, there can be only four different combinations in the offspring of a single pair".[7] Therefore, Sutton suggested that a careful study of the whole process of cell division was necessary. What he found was really important: how the chromosomes of the same pair would be positioned in the equatorial plate of the "reducing division" was entirely a matter of chance. As a result, a large number of different combinations of maternal and paternal chromosomes was possible in reproductive cells.[8] Sutton went on to present calculations of the possible different combinations of chromosomes of an individual, showing that even for small chromosome numbers the number of possible combinations was huge. He thus concluded:

> It is this possibility of so great a number of combinations of maternal and paternal chromosomes in the gametes which serves to bring the chromosome-theory into final relation with the known facts of heredity; for Mendel himself followed out the actual combinations of two and three distinctive characters and found them to be inherited independently of one another and to present a great variety of combinations in the second generation.[9]

Then Sutton immediately proceeded to figure out the implications for the study of heredity. He assumed that an individual had a paternal chromosome A and a maternal chromosome a. During cell division A and a would come together producing the "bivalent chromosome Aa",[10] which during the "reducing division" would be separated into A and a, which would in turn be included in the reproductive cells. Thus, a monoecious plant, that is one having both the male and female reproductive organs, would produce the following four kinds of reproductive cells: female A; male A; female a; male a. These, according to Sutton, would form the following combinations in the offspring:

female A + male A = AA
female A + male a = Aa
female a + male A = aA
female a + male a = aa

Based on this, Sutton concluded:

> Since the second and third of these are alike the result would be expressed by the formula *AA: 2Aa: aa* which is the same as that given for any character in a Mendelian case. Thus the phenomena of germ-cell division and of heredity are seen to have the same essential features, viz., purity of units (chromosomes, characters) and the independent transmission of the same; while as a corollary, it follows in each case that each of the two antagonistic units (chromosomes, characters) is contained by exactly half the gametes produced.[11]

So here were the foundations for explaining the physical basis of Mendel's ratios and for understanding the chromosomal nature of heredity. But Sutton went even further. He assumed that there was "a definite relation between chromosomes and allelomorphs or unit characters" but it was not clear whether a whole chromosome or part of it should be regarded as "the basis of a single allelomorph". Sutton argued for the latter, because otherwise there could not be more distinct characters than the total number of chromosomes. And he concluded:

> We must, therefore, assume that some chromosomes at least are related to a number of different allelomorphs. If then, the chromosomes permanently retain their

individuality, it follows that all the allelomorphs represented by any one chromosome must be inherited together.[12]

Based on his cytological observations, in 1903 Sutton made predictions that would later be confirmed: not only that alleles are found on chromosomes but also that several alleles are found on the same chromosome and that they were therefore linked. Sutton's insights thus set a sufficient basis for a chromosomal theory of heredity.

CHROMOSOME LINKAGE

Even though Sutton followed Bateson's reading of Mendel's work to draw implications from his cytological research for research in heredity, Bateson did not follow Sutton's insights. For Mendel's laws to be applicable, there had to be an independent segregation of characters. Bateson did believe that the study of cell division was key to understanding:

> Stripped of all that is superfluous and of all that is special to particular cases, genetics stand out as the study of the process of cell-division. For if we had any real knowledge of the actual nature of the processes by which a cell divides, the rest would be largely application and extension. It is in cell-division that almost all the phenomena of heredity and variation are accomplished.[13]

Bateson accepted that observations of cytologists were important, yet as there was a lot that was not known, such as why and how phenomena occur the way they do, the study of cells had not yet contributed to a better understanding of heredity. Furthermore, he questioned the importance of chromosomes for heredity:

> If the chromosomes were directly responsible as chief agents in the production of the physical characteristics, surely we should expect to find some degree of correspondence between the differences distinguishing the types, and the visible differences of number or shape distinguishing the chromosomes. So far as I can learn, no indication whatever of such a correspondence has ever been found.[14]

We see therefore that for Bateson there was no indication that chromosomes were the bearers of hereditary properties. He saw no connection between the various inherited characters and any evident differences in the size or shape of the chromosomes; no differences in the chromosomes of the various tissues of the same organism; no connection between the number of chromosomes and the complexity of an organism. One may wonder why Bateson objected to the chromosomal theory of heredity, even after much about it was well understood. Could it be because of the theoretical implications from linkage that went against the independent segregation and assortment of unit factors? Perhaps. Bateson never really accepted the chromosome theory of heredity, rather believing in a theory of heredity and development based on vortices and waves.[15]

It was the influential research of Thomas Hunt Morgan (Figure 7.3) and his students at Columbia University until the early 1920s that would clearly show the relation between factors and chromosomes and the linkage among factors found

FIGURE 7.3 Thomas Hunt Morgan (1866–1945) Alfred F. Huettner Photograph Collection, MBL Archives. Marine Biological Laboratory. Attribution 4.0 International (CC BY 4.0).

on the same chromosome. This research established *Drosophila* (the fruit fly) as a model organism in genetics research. Morgan enrolled for a PhD in 1886 at Johns Hopkins, working under William Keith Brooks (who as we saw very briefly in Chapter 2 had tried to develop a theory of heredity; he had also influenced Bateson). Morgan was likely influenced by Brooks, although he never clearly expressed that. Nevertheless, there were two key differences between the two. First, Morgan was in favor of finding new observations in order to eliminate one of several alternative hypotheses, whereas Brooks simply preferred to support the hypothesis he considered more likely. Second, Morgan preferred to make conclusions that were strongly supported by the facts, relying a lot less on speculation than Brooks.[16]

Interestingly, Morgan rejected both the chromosomal theory of heredity and the idea of linkage as late as 1910:

> Since the number of chromosomes is relatively small and the characters of the individual are very numerous, it follows on the theory that many characters must be contained in the same chromosome. Consequently many characters must Mendelize together. Do the facts conform to this requisite of the hypotheses? It seems to me that they do not. A few characters, it is true, seem to go together, but their number is small, and it is by no means evident that their combination is due to a common chromosome . . . the absence of groupings of characters in Mendelian inheritance seems a fatal objection to the chromosome theory, so long as that theory attempts to locate each character in a special chromosome.[17]

At around the same time, Bateson and Punnett were still struggling with the phenomena they had observed where they had found exceptions from the Mendelian rules. Consider two pairs of allelomorphs, say *A* and *a* as well as *B* and *b*. According to the Mendelian rules, when two *AaBb* individuals were crossed, it was possible to find all kinds of combinations in the offspring: 9 *A_B*:3 *aaB_*:3 *A_bb*:1 *aabb*. Simply put, *A* could be combined with either *B* or *b*, and *B* could be combined with either *A* or *a*—of course, in different proportions. However, Bateson had found two kinds of exceptions to this rule. When one parent had *A* and *B* and the other parent had *a* and *b*, *A* and *B* tended to be found together in the offspring as also did *a* and *b*. Bateson and Punnett described this as "A system of partial coupling under which two factors are generally associated". There was also the opposite case, in which one parent had *A* and *b and* the other parent had *a* and *B*, and *A* and *b* tended to be found together in the offspring as also did *a* and *B*. They described this as "A system of complete repulsion (or as we have sometimes called it, "spurious allelomorphism") under which two factors are never associated in the same gamete".[18] They further added:

> Expressed in a general form, the conclusion to which we have been led is that if A, a, and B, b, are two allelomorphic pairs subject to coupling and repulsion, the factors A and B will repel each other in the gametogenesis of the double heterozygote resulting from the union Ab x aB, but will be coupled in the gametogenesis of the double hetero-zygote resulting from the union AB x ab. The F_1 heterozygote is ostensibly identical in the two cases, but its offspring reveals the distinction.[19]

Morgan considered this explanation, and in the same year he offered an alternative one, which actually stood in contrast to the views he had expressed the previous year about the chromosomal basis of heredity. He wrote:

> In place of attractions, repulsions and orders of precedence, and the elaborate systems of coupling, I venture to suggest a comparatively simple explanation based on results of inheritance of eye color, body color, wing mutations and the sex factor for femaleness in Drosophila. If the materials that represent these factors are contained in the chromo-somes, and if those factors that "couple" be near together in a linear series, then when the parental pairs (in the heterozygote) conjugate like regions will stand opposed.[20]

Morgan added that there was evidence that homologous chromosomes twisted around each other, and when they were separated some materials were more likely to fall together on the same side, whereas others could fall either on the same or on the opposite sides. As a result, coupling was found for some characters but not for others, and this depended "on the linear distance apart of the chromosomal materials that represent the factors". Morgan concluded: *"Instead of random segregation in Mendel's sense we find 'associations of factors' that are located near together in the chromosomes. Cytology furnishes the mechanism that the experimental evidence demands"*.[21]

This was the basis for the conceptualization of the factors as being located sequentially on chromosomes like beads on a string, which Morgan and his colleagues later presented.[22] Linkage has crucial implications for the segregation of factors. How this would occur depended on whether the factors were linked or not. Imagine an individual *AaBb*. During the production of gametes, all combinations are possible because in some cases the gametes are *AB* and *ab*, whereas in others they will be *Ab* and *aB*.

However, when the factors are linked, not all combinations are possible. What matters in this case is which allelomorphs are linked (e.g., whether *A* is linked with *B* and *a* with *b*, or whether *A* is linked with *b* and *a* with *B*). Depending on this linkage, an individual can only have gametes either *AB* and *ab* or *Ab* and *aB*.

Finding numerous mutants in *Drosophila* made the previous naming system used for the factors insufficient. As we saw in Chapter 1, Mendel's characters were binary and so they were simple to describe. When Bateson and his colleagues found more than two different versions of a trait, the presence/absence system was invented to account for the results (see Chapter 6). Yet, even that could not account for the new findings. Therefore, Morgan decided that a new system of naming was required. The solution was to name and describe genes by their location on the chromosomes because this system had an infinite capacity. To achieve this, on March 5, 1912, Morgan asked his students Alfred H. Sturtevant and Calvin Bridges (Figure 7.4) to embark on a project of mapping chromosomes 2 and 3 of *Drosophila melanogaster.*[23]

The key paper on this topic was published in 1913 by Sturtevant (Figures 7.4 and 7.5), one of Morgan's students and close collaborators. That paper contained the first genetic map, that is, the first map showing the linear arrangement of genes on chromosomes. There, Sturtevant also set out the logic for genetic mapping. This was based on the phenomenon of crossing over: the exchange of chromosome parts between two homologous chromosomes during meiosis, which results in new combinations of factors in offspring that did not exist in their parents. Sturtevant explained

FIGURE 7.4 Morgan's students in the lab. From left to right: Calvin Bridges (1889–1938), Alfred H. Sturtevant (1891–1970), with O.L. Mohr in 1920.

FIGURE 7.5   Morgan in the Fly Room at Columbia in 1917. The person on the right is Alfred H. Sturtevant, who took the photo and titled it "The Boss". According to Rubin and Lewis (2000), Morgan was camera shy, and so Sturtevant took the photo by using a camera hidden in an incubator and operated remotely with a string. Courtesy of the Archives, California Institute of Technology.

that the number of crossovers per 100 cases was used as an index of the distance between any two factors. If one could thus determine the distances between factors A and B and between factors B and C, it would also be able to predict the distance between factors A and C. Therefore, the relative positions of factors could be empirically mapped on chromosomes, and this gave them a more material character than ever before.[24] It should be noted here that this understanding became possible in part because of their work with *Drosophila* that has four chromosomes only, which makes the identification of linked factors more likely.

FROM THE MENDELIAN "FACTORS" TO THE "GENE" CONCEPT

In 1909 Morgan wrote an article titled "What Are Factors in Mendelian Explanations?" that is worth quoting at length:

> In the modern interpretation of Mendelism, facts are being transformed into factors at a rapid rate. If one factor will not explain the facts, then two are invoked; if two prove insufficient, three will sometimes work out. The superior jugglery sometimes

necessary to account for the results may blind us, if taken too naively, to the commonplace that the results are often so excellently "explained" because the explanation was invented to explain them. We work backwards from the facts to the factors, and then, presto! explain the facts by the very factors that we invented to account for them. I am not unappreciative of the distinct advantages that this method has in handling the facts. I realize how valuable it has been to us to be able to marshal our results under a few simple assumptions, yet I cannot but fear that we are rapidly developing a sort of Mendelian ritual by which to explain the extraordinary facts of alternative inheritance. So long as we do not lose sight of the purely arbitrary and formal nature of our formulae, little harm will be done; and it is only fair to state that those who are doing the actual work of progress along Mendelian lines are aware of the hypothetical nature of the factor-assumption. But those who know the results at second hand and hear the explanations given almost invariably in terms of factors, are likely to exaggerate the importance of the interpretations and to minimize the importance of the facts.[25]

We should keep the last sentence in mind for the chapters that follow. The overemphasis on factors exaggerates their importance and shifts attention from the facts to their interpretation. Morgan criticized the overly simplistic accounts of Mendelian genetics, which, as we saw, was promoted by Bateson (and many others). He was very critical of the fact that a simple explanation was taken for granted without considering that it was invented to explain the facts. The "factors" became very important to the extent that, Morgan subsequently noted, they were referred to as being the actual characters themselves—"unit-characters". He noted that this was a form of preformation, the—according to him unfruitful—idea that an individual starts in some already preformed or predetermined way. This has been contrasted to another idea, epigenesis, according to which the individual starts from unformed material and the form emerges gradually, over time.[26]

Referring to the inheritance of eye color in mice, Morgan noted that in heterozygotes the same character could in some cases be the dominant and in other cases the recessive. Moreover, in the second generation, a continuous series of types had often been found, showing that they were not in agreement with the Mendelian rules of the segregation of characters. Then Morgan provided an account of how heredity occurred, which would be more accurate if we taught it at schools even today instead of the naive account of Mendelian genetics that I described in Chapter 1:

> The egg need not contain the characters of the adult, nor need the sperm. Each contains a particular material which in the course of the development produces in some unknown way the character of the adult. Tallness, for instance, need not be thought of as represented by that character in the egg, but the material of the egg is such that placed in a favorable medium it continues to develop until a tall plant results. Similarly for shortness.[27]

Morgan had been an experimental embryologist and therefore sensitive to the complexities of development. In this quotation he clearly suggested, again as we should be doing today in schools, that the hereditary material is just a propensity toward an outcome that also depends on the surrounding conditions; in other words, the outcome is not predetermined therein. He therefore argued that referring to the materials in the reproductive cells as if they were one and the same thing as the adult characters was not justified, because alternative outcomes were possible. He also noted that, as it was well known, Mendelian results were only average results.

It is perhaps strange to read such a criticism by someone who would soon develop an exemplary model for Mendelian genetics and who would later be awarded the Nobel Prize for this work. Morgan initially rejected both Mendelian genetics and the chromosome theory of heredity. He also rejected Darwin's view that species evolve by gradual change through natural selection. He was rather fascinated by the idea of discontinuous variation and by de Vries' idea that species evolve through discrete jumps, called mutations (also Darwin's and Galton's "sports"). This made him in 1908 to start experiments with *D. melanogaster* in an attempt to find such mutations. Along with Fernandus Payne, Morgan investigated the effect of various factors, such as X-ray radiation, centrifugation, acids, bases, salts, and others, on *Drosophila*, following de Vries' theory that mutation was induced by changes in the environment and was thus adaptive and responsive. For Morgan, evolution occurred via discontinuous mutations that nevertheless fell within the normal range of variation.[28] It was Morgan's work with *Drosophila* in what he called sex-limited inheritance, and the evidence published during the decade 1900–1910, that made him eventually accept both the Mendelian theory and the chromosome theory as generally applicable to hereditary phenomena.[29]

In 1909, Wilhelm Johannsen (Figure 7.6) proposed the term "gene" to refer to these hereditary factors, in his book *Elemente der exakten Erblichtkeitslehre*

FIGURE 7.6   Wilhelm Johannsen (1857–1927) with William Bateson seated in the garden at Merton Park (John Innes Conservation Area). John Innes Archives courtesy of the John Innes Foundation.

(*Elements of the Exact Theory of Heredity*). Etymologically, the term derived from de Vries' "Pangens" the hereditary factors in his intracellular pangenesis theory (see Chapter 5), which were sometimes transcribed as "Pangenes", with the idea, of course, going back to Darwin's Pangenesis. Johannsen suggested that there was "something" in the gametes or in the zygote that was essential for the characters of the organism. Yet, it was usually described with the rather ambiguous term "pre-disposition" or others that had been introduced in accordance with particular hypo-thetical views. The word "pangene", which he attributed to Darwin, was a good alternative for him but only partially. After analyzing the etymology of the word, he noted that only the second part that meant "becoming" ("γί- γ(ε)ν- ομAι") was of interest, in order to express the simple idea that a property of the developing organism is or can be conditioned or co-determined by something in the gametes. He therefore suggested:

> Therefore, it seems easiest to use the last syllable "gene" isolated from Darwin's known word, which is the only syllable of interest to us, in order to replace the bad, ambigu-ous word "disposition". . . . The word "gene" is completely free of any hypothesis; it expresses only the certain fact that in any case many properties of the organism are conditioned by special, separable and thus independent "states", "foundations", "dispo-sitions" occurring in the gametes—in short, what we want to call genes.[30]

For Johannsen, the gene concept was free from any assumption about its localization in the cell and its material constitution. In the same book, Johannsen also introduced the concepts of genotype and phenotype. He defined again all three key terms in a 1911 article:

> The "gene "is nothing but a very applicable little word, easily combined with others, and hence it may be useful as an expression for the "unit-factors", "elements" or "allelo-morphs" in the gametes, demonstrated by modern Mendelian researchers. A "genotype" is the sum total of all the "genes" in a gamete or in a zygote. When a monohybrid is formed by cross fertilization, the "genotype" of this F1-organism is heterozygotic in one single point and the "genotypes" of the two "genodifferent" gametes in question differ in one single point from each other. As to the nature of the "genes" it is as yet of no value to propose any hypothesis; but that the notion "gene" covers a reality is evident from Mendelism. . . . All "types" of organisms, distinguishable by direct inspection or only by finer methods of measuring or description, may be characterized as "phenotypes".[31]

Morgan did not adopt the concept of "gene" immediately. In 1913, he tried to further clarify the relation between "unit characters" and "unit factors". He wrote:

> The factorial hypothesis has played an important role in Mendelian heredity, and while students of Mendel's principles have had on the whole a pretty clear idea of the sense or senses in which they have made use of factors or symbols, yet those not engaged in the immediate work itself have, I believe, often been misled in regard to the meaning attached to the term factor, and by the assumed relation between a factor and a unit character. The confusion is due to a tendency, sometimes unintentional, to speak of a unit character as the product of a particular unit factor acting alone, but this identifica-tion has no real basis.[32]

So already in 1913, Morgan noted that thinking of a character as the product of a particular factor was due to persistent confusion and had no real basis. This is why he wanted to clarify his own view. Morgan then explained that to think that one factor alone could determine a character was misleading and stemmed from a misunderstanding of the notations used. Whereas the notation scheme commonly used could account for only two versions of the same character, it was already clear that several different versions—more than two—of the same character had been found, as was the case for body color or wing shape in *Drosophila*.

In 1915, in a book written by Morgan and his students and collaborators Alfred Sturtevant, Calvin Bridges, and Hermann J. Muller (Figure 7.7), they introduced the notion of factors (later genes) "for" a character or trait. For instance, they wrote:

> As shown in this diagram, a spermatozoon bearing the factor for long wings fertilizing an egg bearing the same factor produces a fly pure for long wings; a spermatozoon bearing the factor for long wings fertilizing an egg bearing the factor for vestigial wings produces a hybrid fly that has long wings. Hence we say the long dominates the vestigial character. Similarly, a spermatozoon bearing the factor for vestigial wings fertilizing an egg bearing the factor for long wings produces a hybrid with long wings; a spermatozoon bearing the factor for vestigial wings.[33]

FIGURE 7.7   Herman J. Muller (1890–1967) in 1927. Marine Biological Laboratory Black Photo Album 2. Attribution 4.0 International (CC BY 4.0).

This way of referring to "factors for characters" is used throughout that book, and we read about "the factor of ebony",[34] "the factor for red",[35] and so on. However, they also explained that whereas it was "customary to speak of a particular character as the product of a single factor", everyone familiar with the phenomena of Mendelian inheritance was aware that the "so-called unit character is only the most obvious or most significant product of the postulated factor".[36]

In the same book, Morgan and his colleagues wrote something that I consider the essence of how we should be thinking and talking about genes today.

> Mendelian heredity has taught us that the germ cells must contain many factors that affect the same character. Red eye color in Drosophila, for example, must be due to a large number of factors, for as many as 25 mutations for eye color at different loci have already come to light. . . . Each such color may be the product of 25 factors (probably of many more) and each set of 25 or more differs from the normal in a different factor. It is this one different factor that we regard as the "unit factor" for this particular effect, but obviously it is only one of the 25 unit factors that are producing this effect. . . . The converse relation is also true, namely, that a single factor may affect more than one character. . . . Failure to realize the importance of these two points, namely, that a single factor may have several effects, and that a single character may depend on many factors, has led to much confusion.[37]

This is the important distinction between factors as character-makers and factors as character-difference-makers. It is one thing to suggest that a factor makes a character and another that a factor makes a difference in the state of a character. This way of describing the role of genes conveys two messages: (1) that between factors and characters there is a many-to-many and not a one-to-one relationship; and (2) that, this notwithstanding, it was possible that particular changes in particular factors could bring about changes in particular characters. This is what a factor "for" a character meant. It was just a shorthand to refer to a difference-maker, and in no way did it mean that a factor could produce on its own a whole character. In today's parlance, already in 1915, it was clear to Morgan and his collaborators that traits are polygenic and factors/genes are pleiotropic, even though changes in some of them might occasionally bring about more significant differences than changes in others.

The comparison between the last two quotations is both interesting and instructive. The first one is an exemplar representation of determinism, the view that single factors determine characters, which is why they are described as the factors "for" the respective character. This is a way that not only references to genes and traits and their relationship is made even today but also one that is often interpreted literally. This is also what the descriptions by Bateson and his colleagues implied. However, Morgan and his collaborators were well aware, already in 1915, that this was not how things worked. They thus explained that the notion of a factor "for" a trait was just a shorthand for a more complex relationship, the one described in the second quotation. This is a clear representation of an anti-determinist view, according to which a factor can make a difference in the occurrence of an effect, which is what being "for" something means, but at the same time is only one of the various factors (genetic and environmental) implicated in the development of the character. It is the latter concept that we need to teach today in schools; however, more often than not teaching tends to imply the former. Why we still do not teach what was clear to Morgan and his collaborators already in 1915 puzzles me.

Morgan eventually adopted the concept of "gene" in 1917. He wrote:

> The germ plasm must, therefore, be made up of independent elements of some kind. It is these elements that we call genetic factors or more briefly genes. This evidence teaches us nothing further about the nature of the postulated genes, or of their location in the germ plasm.[38]

Then Morgan made the interesting remark: "Why, it may be asked, is it not simpler to deal with the characters themselves, as in fact Mendel did, rather than introduce an imaginary entity, the gene".[39] This was Morgan interpreting Mendel (correctly, as I have shown in Chapter 1) as focusing on characters, and he asked why one would need to take a step further and invoke an imaginary entity. I honestly wonder what those scholars who have suggested that Mendel thought in terms of hereditary particles would have to say about why Morgan wrote this. Nevertheless, he defended the need to use the concept of genes, even if there was no evidence about their nature and localization. This was done for heuristic purposes that had to do with explaining the phenomena observed, as the gene concept had already been a very valuable heuristic tool for conducting research.

In 1926, in his book *The Theory of the Gene*, Morgan presented all the available evidence showing that genes were located on chromosomes. He described this theory as being based on particular principles: (1) that the paired genes related to particular characters in an individual; (2) that the genes of the same pair separated in accordance with Mendel's first law; (3) that genes that were not linked assorted independently in accordance with Mendel's second law; (4) that crossing over took place; and (5) that crossing over provided evidence for the linear order of genes and their relative positions.[40] This theory was based on particular assumptions: that genes were relatively constant, that they could be multiplied, and that they were united and then separated during the maturation of the reproductive cells.[41] In the same book, Morgan also argued that the most complete and convincing evidence for the importance of chromosomes in heredity came from research on the effects of particular changes in the number of chromosomes. For instance, flies with only one chromosome 4 developed to be slightly different in many parts of the body compared to the diploid ones. This was interpreted as showing that the presence of a single chromosome 4 was not adequate to produce a "normal" phenotype and that therefore some crucial genes should be located on the missing one.[42] Morgan concluded the book with a section titled: "Are Genes of the Order of Organic Molecules?" His conclusion was that this question was difficult to answer. However, he estimated that genes should have the size of the larger organic molecules and that they were constant exactly because they were chemical entities.[43]

In 1933, Morgan received the Nobel Prize in Physiology and Medicine "for his discoveries concerning the role played by the chromosome in heredity". In his Nobel lecture, he emphasized that it would be "somewhat hazardous to apply only the simpler rules of Mendelian inheritance; for, the development of many inherited characters depends both on the presence of modifying factors and on the external environment for their expression". In other words, Morgan suggested that the simple Mendelian genetics could not account for the development of phenotypes as "the gene generally produces more than one visible effect on the individual, and . . . there may be also many invisible effects of the same gene". The view that genes affect but do not determine characters continued to characterize his thinking. In the same lecture Morgan also clearly explained why at that point it did not make much difference for geneticists to be aware of what genes were made of:

> There is no consensus of opinion amongst geneticists as to what the genes are whether they are real or purely fictitious—because at the level at which the genetic experiments lie, it does not make the slightest difference whether the gene is a hypothetical unit, or whether the gene is a material particle. In either case the unit is associated with a specific chromosome, and can be localized there by purely genetic analysis. Hence, if the gene is a material unit, it is a piece of a chromosome; if it is a fictitious unit, it must be referred to a definite location in a chromosome—the same place as on the other hypothesis. Therefore, it makes no difference in the actual work in genetics which point of view is taken.[44]

But others thought not only that genes were real but also that they determined who and what people were.

It was in the same year, 1933, that the Nazis came to power in Germany and began immediately to discriminate among people using biology as a criterion and justification. But they did not come out of nowhere. In fact, they were largely influenced by the developments in Britain, and particularly in the United States, of a movement that came to be known as "eugenics". A key feature of this movement in the United States and in Nazi Germany was the assumption that complex mental human traits could be due to single, or just a few, recessive genes—an idea that I henceforth describe as "Mendelian" eugenics.[45] This movement emerged in the United States around 1909, and subsequently in Germany, despite the work of Morgan and others who had clearly shown the complexities of heredity and development.

NOTES

1 Allen (1975), p. 42.
2 Tschermak (1951), p. 170.
3 Harris (1999), pp. 170–172; Maderspacher (2008).
4 Sutton (1902), p. 39.
5 Sutton (1903), p. 231.
6 Sutton (1903), p. 231.
7 Sutton (1903), p. 233.
8 Sutton (1903), pp. 233–234.
9 Sutton (1903), p. 235.
10 Like other people at the time, Sutton thought that paternal and maternal chromosomes joined each other before cell division and that their separation and independent assortment took place during the second meiotic division. It took years before how meiosis actually occurs was figured out. See Hegreness and Meselson (2007).
11 Sutton (1903), p. 237.
12 Sutton (1903), p. 240.

13 Bateson (1907), p. 650.
14 Bateson (1907), pp. 651–652.
15 Rushton (2014).
16 Allen (1978), Chapter 2.
17 Morgan (1910), pp. 467–468.
18 Bateson and Punnett (1911), p. 3.
19 Bateson and Punnett (1911), p. 4.
20 Morgan (1911), p. 384.
21 Morgan (1911), p. 384, emphasis in the original.
22 Morgan et al. (1915), pp. 131–132.
23 Kohler (1994), pp. 58–62.
24 Sturtevant (1913).
25 Morgan (1909), p. 365.
26 Maienschein (2021).
27 Morgan (1909), p. 367.
28 Kohler (1994), p. 41.
29 Allen (1978), pp. 125, 144–145; Kohler (1994), pp. 37–39.
30 Johannsen (1909), p. 124 (author's translation); see also Roll-Hansen (2014), p. 1011.
31 Johannsen (1911), pp. 133–134.
32 Morgan (1913), p. 5.
33 Morgan et al., 1915, pp. 11–12.
34 Morgan et al., 1915, p. 13.
35 Morgan et al., 1915, p. 18.
36 Morgan et al. (1915), p. 32.
37 Morgan et al. (1915), pp. 208–210.
38 Morgan (1917), p. 515.
39 Morgan (1917), p. 517.
40 Morgan (1926), p. 25.
41 Morgan (1926), p. 27.
42 Morgan (1926), pp. 45–48.
43 Morgan (1926), pp. 309–310.
44 www.nobelprize.org/prizes/medicine/1933/morgan/facts/ (accessed May 28, 2023).
45 In fact, it would more accurately be described as "Batesonian" eugenics as what we call
   Mendelian genetics today were actually Bateson's interpretation of Mendel's paper.

8 "Mendelian" Eugenics

GALTON'S DREAM

In his book *Memories of My Life*, Francis Galton paid tribute to Mendel's achievement by writing:

> I must stop for a moment to pay a tribute to the memory of Mendel, with whom I sentimentally feel myself connected, owing to our having been born in the same year 1822. His careful and long-continued experiments show how much can be performed by those who, like him and Charles Darwin, never or hardly ever leave their homes, and again how much might be done in a fixed laboratory after a uniform tradition of work had been established. Mendel clearly showed that there were such things as alternative atomic characters of equal potency in descent. How far characters generally may be due to simple, or to molecular characters more or less correlated together, has yet to be discovered.[1]

As I mentioned in the Preface, in 2022 there were several celebrations for the bicentenary of Mendel's birth. But there was nothing like that about Galton, despite his seminal contributions to statistics, forensics, and other fields (by the way, Louis Pasteur was also born in 1822). But this is not all. In 2020, University College London (UCL) announced that the Galton Lecture Theatre, along with the Pearson Building and the Pearson Lecture Theatre, would be renamed. The Lecture Theatre had been named after Galton, because upon his death in January 1911, UCL received 45,000 pounds, an amount sufficient to provide 1,500 pounds a year for the establishment of a Galton Eugenics Professorship. In accordance with Galton's wish, the post was offered to Karl Pearson.[2] But because of their involvement in the eugenics movement, maintaining places in UCL named after them did not seem appropriate. As UCL's President and Provost Professor Michael Arthur put it:

> This is an important first step for UCL as we acknowledge and address the university's historical links with the eugenics movement. . . . Although UCL is a very different place than it was in the 19th century, any suggestion that we celebrate these ideas or the figures behind them creates an unwelcoming environment for many in our community.[3]

I personally do not have any strong views for or against such decisions. However, I am wondering how far such practices can, or ought to, go. If we reject Galton and Pearson, should we also refrain from using their statistical methods? And how should we treat other scientific figures who held unacceptable views? For instance, Darwin could be considered a racist in some sense, given that he considered non-Europeans as inferior to Europeans, although he despised slavery.[4] He can also be considered sexist, given that he considered women as having a lower intellect than men, although he helped several women naturalists get published.[5] In the *Descent of Man*, he had written—citing Galton among others—that civilized societies tend to counter

DOI: 10.1201/9781032449067-10

the eliminative action of natural selection against "the weak in body or mind". He explained that:

> we build asylums for the imbecile, the maimed, and the sick; we institute poor-laws; and our medical men exert their utmost skill to save the life of every one to the last moment. . . . Thus the weak members of civilized societies propagate their kind. No one who has attended to the breeding of domestic animals will doubt that this must be highly injurious to the race of man.[6]

Hardly any breeder would thus allow the worst animals to breed, given that this would lead to the degeneration of a domestic race. But humans made an exception and did this to themselves. Darwin was nevertheless in favor of supporting the weak. He thus added:

> Nor could we check our sympathy, if so urged by hard reason, without deterioration in the noblest part of our nature. . . . Hence we must bear without complaining the undoubtedly bad effects of the weak surviving and propagating their kind; but there appears to be at least one check in steady action, namely the weaker and inferior members of society not marrying so freely as the sound; and this check might be indefinitely increased, though this is more to be hoped for than expected, by the weak in body or mind refraining from marriage.[7]

Is this a proto-eugenic view? If yes, should we dethrone Darwin too? I am inclined to think that such decisions are a bit arbitrary and local.

The foundations of eugenics were set by Galton. He expressed his ideas about improving mankind for the first time in his 1865 paper "Hereditary Talent and Character". There he wondered "how vastly would the offspring be improved" if distinguished women were commonly married to distinguished men, according to rules yet unknown but there to be revealed.[8] His conclusion?

> I hence conclude that the improvement of the breed of mankind is no insuperable difficulty. If everybody were to agree on the improvement of the race of man being a matter of the very utmost importance, and if the theory of the hereditary transmission of qualities in men was as thoroughly understood as it is in the case of our domestic animals, I see no absurdity in supposing that, in some way or other, the improvement would be carried into effect.[9]

In 1869, he proposed the idea of selective breeding among humans with the aim of improving the human race as a whole, in his book *Hereditary Genius*:

> What I profess to prove is this: that if two children are taken, of whom one has a parent exceptionally gifted in a high degree—say as one in 4,000, or as one in a million—and the other has not, the former child has an enormously greater chance of turning out to be gifted in a high degree, than the other. Also, I argue that, as a new race can be obtained in animals and plants, and can be raised to so great a degree of purity that it will maintain itself, with moderate care in preventing the more faulty members of the flock from breeding, so a race of gifted men might be obtained, under exactly similar conditions.[10]

But it was not until 1883 that Galton gave this project the name "eugenics". He coined the term from the Greek word "eugenes" (ευγενής), meaning "good in stock, hereditary endowed with noble qualities". He continued:

> We greatly want a brief word to express the science of improving stock, which is by no means confined to questions of judicious mating, but which, especially in the case of man, takes cognizance of all influences that tend in however remote a degree to give to the more suitable races or strains of blood a better chance of prevailing speedily over the less suitable than they otherwise would have had.[11]

It is now well known that eugenic policies were implemented in various countries during the first half of the 20th century. Britain, the birthplace of the idea, was not one of them. In the present chapter we consider what happened in the United States and in Chapter 10 what happened in Germany, but there were many other countries, mostly in Europe but also in Canada, in which such policies were implemented. In all cases, this implementation was characterized by biases and discrimination, that varied from sterilization in the United States to the extermination of millions of people in Nazi Germany. Generally speaking, there have been two types of eugenics (a distinction first made by physician and eugenicist Caleb Williams Saleeby).[12] On the one hand, what has been described as "positive eugenics" was about encouraging the "fittest" or most superior individuals to reproduce early and often. On the other hand, what has been described as "negative eugenics" is aimed at discouraging the "less fit" or "unfit" members of society from having children, something often achieved through their sterilization. It is indeed difficult to imagine one without the other. But much as Galton has been blamed for these ideas, one can see in his writings a focus on the former rather than the latter. In 1904, Galton founded the National Eugenics Laboratory, which was followed by the foundation of the Eugenics Education Society in 1907 and of the journal *Eugenics Review* in 1909.

In 1904 Galton read a paper titled "Eugenics: Its Definition, Scope, and Aims" before the Sociological Society at a meeting in London, on May 16, with Pearson chairing the session. Galton defined eugenics as:

> the science which deals with all influences that improve the inborn qualities of a race; also with those that develop them to the utmost advantage. The improvement of the inborn qualities, or stock, of some one human population will alone be discussed here.[13]

When Galton described the aim of eugenics, he focused on the positive aspect: "The aim of eugenics is to bring as many influences as can be reasonably employed, to cause the useful classes in the community to contribute *more* than their proportion to the next generation".[14] The aim for him was to make the "superior" individuals contribute more offspring to the next generation so that their overall proportion increased.

Galton's focus on the positive version of eugenics did not convince some of the discussants. For instance, the writer Herbert George Wells commented:

> The fact that the sons and nephews of a distinguished judge or great scientific man are themselves eminent judges or successful scientific men may after all, be far more due to a special knowledge of the channels of professional advancement than to any

distinctive family gift. I must confess that much of Dr. Galton's classical work in this direction seems to me to be premature.

Wells instead suggested that:

The way of nature has always been to slay the hindmost, and there is still no other way, unless we can prevent those who would become the hindmost being born. It is in the sterilization of failures, and not in the selection of successes for breeding, that the possibility of an improvement of the human stock lies.[15]

For Wells positive eugenics seemed complicated; there was not sufficient knowledge about what to select and how, whereas negative eugenics seemed straightforward to him: sterilize "failures" so that they would not reproduce. Galton concluded the article by noting that even though improving the human stock seemed to him one of the highest aims, "Overzeal leading to hasty action would do harm, by holding out expectations of a near golden age, which will certainly be falsified and cause the science to be discredited".[16] Unfortunately, and ironically, what Galton was right about was that eugenics would do harm.

Bateson did not attend the meeting but later submitted a written comment. He expressed his sympathy to Galton's paper and suggested that without a doubt this discussion would "do something to promote the study of heredity and the introduction of scientific method in the breeding of man and other animals". He concluded: "I would, therefore, urge that those who really have such aims at heart will best further '*eugenics*' by promoting the attainment of that solid and irrefragable knowledge of the physiology of heredity which experimental breeding can alone supply".[17] Weldon attended the meeting. He was not enthusiastic about eugenics, which he considered premature, but wanted to seize the opportunity to criticize the Mendelians.[18] He first stated that "a large number of apparently simple results have been attained in experimental breeding" by Mendel, and which were later carried out by Bateson and de Vries with a focus on simple characters. Then he argued that:

from a laboratory experiment you have not arrived at the formulation of a eugenic maxim. You must look at your facts in their relation to an enormous mass of other matter, and in order to do that you must treat large masses of your race in successive generations, and you must see whether the behavior of these large masses is such as you would expect from your limited experiment. If the two things agree, you have realized as much of the truth as would serve as a basis for generalization. But if you find there is a contradiction resulting from the facts—from the large masses and limited laboratory experiments—then there is no doubt whatever that, from the point of view of human eugenics, and from the theory of evolution, the more important data are those from the larger series of material; the less important are those from laboratory experiment.[19]

Whatever Weldon's motivations for this comment were, he raised a very important issue. That experiments in controlled environments that focused on simple characters and gave simple results had limitations. Generalizations were possible only when one had evidence from large populations. In other words, whatever worked in an experimental setting might not be confirmed from a larger series of data, and one then should be ready to discard the respective experiment. If Mendelian genetics worked for peas in the laboratory, would they work for peas in nature? And if they did not, why would they work for humans?

Weldon's close friend, and chair of that session, Karl Pearson (Figure 8.1) succeeded Galton as the main eugenicist in Britain. He had gone to Germany for studies of law, philosophy, and mathematics. Whereas he disliked Germany of the time and called the Prussians "barbaric", he was attracted to German idealism. While there, his first name had been changed from Carl to the German equivalent "Karl". When in 1880 he returned to England, he maintained "Karl" and dreamed of having a German wife. He also accepted the views of Johann Fichte that citizens should be subordinate to the welfare of the state. In 1890, he eventually married Maria Sharpe, an intellectual woman, who made Pearson concede that women's lack of intellectual achievement was due to the lack of opportunities. Pearson made important contributions to mathematics during his close collaboration with Weldon while he was also teaching at UCL.[20] They jointly battled Bateson and the Mendelians in the 1900s, and Weldon's death in 1906 devastated Pearson. The memoir that Pearson wrote for him in 1906, first published in *Biometrika*, the journal they had co-founded, began like this:

> It is difficult to express adequately the great loss to science, the terrible blow to biometry, which results from the sudden death during the Easter vacation of the joint founder,

FIGURE 8.1   Karl Pearson (1857–1936). Wellcome Collection. Attribution 4.0 International (CC BY 4.0)

and co-editor of this journal. The difficulty of adequate expression is the greater, because so much of Weldon's influence and work were of a personal character.[21]

Pearson became interested in eugenics around 1900. He was a lot more competent than Galton in mathematics, and he was instrumental in developing key statistical methods for the analysis of data. In 1907, he became the supervisor of the Francis Galton Laboratory for National Eugenics at UCL, working in consultation with Galton. As already mentioned, upon Galton's death in 1911, Pearson became Galton Professor of Eugenics at UCL and focused on eugenic projects. This relieved Pearson from his heavy teaching duties and gave him time for research. He relied extensively on volunteers for the collection of data, all in various places in England. This allowed for the study of the variability of populations and the correlations found among relatives for various traits and disorders. The outcome was dozens of studies, many of which were published in Pearson's own journal *Biometrika*.[22] According to historian Joe Cain, Pearson was a racist but not in the sense of white supremacy over other human races we usually encounter. Rather, Pearson was concerned about the local "stock" in his country, which was Anglo-Saxon, and whether it might be deteriorated. As Cain put it, Pearson was "a racist, nativist, supremacist for an imaginary Anglo-Saxon 'stock' ".[23]

Pearson's views are clearly illustrated in a 1925 co-authored paper with the title "The Problem of Alien Immigration into Great Britain, Illustrated by an Examination of Russian and Polish Jewish Children". There he wrote:

> The student of national eugenics desires in every way to improve and strengthen his own nation. He would do this by intra-national selection for parentage, and by the admission wherever and whenever possible of superior brains and muscles into his own country.

The goal of the project is clear here. To improve the "stock" of the nation, the new generations had to emerge from unions between natives. If foreigners were to be allowed to contribute to the national stock, they had to be of superior quality. But then, Pearson continued, not as much as required was done to allow this. "We simply have not the knowledge at present requisite to set a value on the hybrids which may result from crosses even of the physically and mentally best in Caucasian, Oriental and African races".[24] This was a problem for both immigration and emigration, he noted.

An important concern for Pearson was the problem of "indiscriminate immigration" of Polish and Russian Jews into the London East End, as well as poor areas in other towns in the country, before World War I. Whereas the immigrants were as hard-working as the English, they had lower standards of quality of life and contributed to the spread of doctrines that destabilized the social and political institutions. The wish for cheap labor facilitated their entrance, and not much thought had been given as to whether the nation would benefit from assimilating "this mass of new and untested material". This was, according to Pearson, why the views of eugenicists had to be taken into account:

> The whole problem of immigration is fundamental for the rational teaching of national eugenics. What purpose would there be in endeavouring to legislate for a superior breed of men, if at any moment it could be swamped by the influx of immigrants of an

inferior race, hastening to profit by the higher civilisation of an improved humanity? To the eugenist permission for indiscriminate immigration is and must be destructive of all true progress. Such progress is only possible where intra-racial selection is combined with a large measure of isolation.[25]

In short, for Pearson indiscriminate immigration would be destructive; he considered it imperative to avoid the influx of inferior material, if the goal was for the nation to reach a superior level. Pearson added though that in the past some special cases of immigrants of "high mentality and firm purpose" had been beneficial.

A common concern among eugenicists was how to deal with the so-called "feeblemindedness". This is the case I focus on in this and the next chapter of the present book (also always in quotation marks because it was neither a single condition nor a clearly defined one). During the last decade of the 19th century, a new problematic social group emerged: high-grade mentally defective children, initially, and subsequently adolescents and adults. Proposals about how to deal with this "problem" were suggested by the *Report of the Royal Commission on the Care and Control of the Feeble-Minded*, which was published in 1908 in Britain. One of the main findings of the report was that the mentally defective were involved in various social problems, including crime, prostitution, alcoholism, or vagrancy. They were also mostly unemployed and dependent, as well as sexually promiscuous, thus spreading venereal disease. Therefore, if the mentally defective regularly transferred their defects to the next generations, something had to be done about it.[26]

However, Britain never legislated for, or carried out, compulsory sterilizations. Since the Mental Deficiency Act of 1913, only detention in institutions had taken place. This becomes even more impressive if one considers that prominent politicians such as Winston Churchill held racist views and had eugenics aspirations. As his official biographer has suggested, as early as 1899, Churchill had written to a cousin of him that "The improvement of the British breed is my aim in life". When he entered the Parliament in 1901, Churchill considered the so-called feebleminded and the insane as a threat to British society. Looking at the respective legislation in the United States, Churchill believed that sterilization and segregation of those people was the solution. During the time he was home secretary (February 1910–October 1911), Churchill was in favor of the confinement, segregation, and sterilization of people contemporarily described as the "feebleminded".[27]

But the situation on the other side of the Atlantic was very much different.

"FEEBLEMINDEDNESS" BECOMES A MENDELIAN RECESSIVE TRAIT

The person who perhaps did more than anyone else to advance eugenic ideals in the United States (and elsewhere, as we soon see) was Charles Davenport (Figure 8.2). Having initially studied engineering, he had acquired important skills in mathematics. While he was an instructor in zoology at Harvard during the 1890s, he became an important pioneer in biometry. He had met Galton and Pearson in England and was influenced by them. In 1899 he left Harvard for a post at the University of Chicago. Following the trends of the time that favored a focus on experimentation, he persuaded the Carnegie Institution in Washington to fund a station for the experimental study of evolution. This was established in 1904 at Cold Spring Harbor, with Davenport as its director. The focus initially was on the study of inheritance in poultry and canary; however, attention was soon turned to human traits such as eye, hair, and skin color. Davenport became eager to study more human traits. As experimental work was, of course, impossible, he decided to accumulate systematic and detailed data on human pedigrees. Whenever he found a high incidence of a character, Davenport concluded that it must be inheritable and tried to account for it via Mendelian schemes that were often oversimplified. To accumulate as much data as possible, he started looking for funds. He therefore approached Mary Harriman, who had spent a summer as an undergraduate at Cold Spring Harbor for support. Mary was the daughter of Edward Henry Harriman, a railroad tycoon, who had died in 1909 and whose widow was managing his immense fortune. Mary helped Davenport convince her mother to fund the Eugenics Record Office, which was founded in 1910 next to the Cold Spring Harbor Experimental station. The money was used to hire staff who were trained and then sent away to accumulate data from families all over the United States. Harry Laughlin (Figure 8.2) was appointed as the superintendent of the Eugenics Record Office, a post he held until its closure in 1939.[28]

Despite having met Galton and Pearson very early on in his career, Davenport was not interested in eugenics right from the start. Historian Philip Pauly has argued

FIGURE 8.2 From right to left, Charles Benedict Davenport (1866–1944) and Harry Laughlin (1880–1943) outside the new Eugenics Record Office building at Cold Spring Harbor, around 1913. Courtesy of Harry H. Laughlin Papers, Special Collections & Museums Department, Pickler Memorial Library, Truman State University.

that Davenport embraced the eugenics movement through his involvement in the American Breeders' Association, the first national association to promote eugenics research, as secretary of its committee on eugenics. Another reason was Davenport's close interaction with Theodore Roosevelt, President of the United States from 1901 to 1909, who believed that white people of European descent were innately superior to all other human races, and who thus endorsed eugenic views.[29] In a 1909 paper, Davenport endorsed both negative and positive eugenics. He noted that "First, the scandal of illegitimate reproduction among imbeciles must be prevented. That class often shows a frightful fecundity". If segregation was insufficient, and if the "destruction of the lowest-grade imbeciles" was not acceptable, it was required to find a way to "prevent the reproduction of their vicious germ plasm". Second, he noted that there were no classes superior to others; it was all about individuals.

> The way to improve the race is first to get facts as to the inheritance of different characteristics and then by acquainting people with the facts lead them to make for themselves suitable matings. The only rule, a very general one, that can be given at present is that a person should select as consort one who is strong in those desirable characters in which he is himself weak, but may be weak where he is strong.[30]

It was also in 1909 that Davenport started a close correspondence and interaction with the psychologist Henry Goddard, Director of Research at the Vineland Training School for the Feeble-Minded in New Jersey (Figure 8.3). In 1908, Goddard

FIGURE 8.3 Henry Herbert Goddard (1866–1957) in 1910, when he was writing his major works on "feeblemindedness". Public Domain. Wikimedia Commons.

introduced in the United States the tests developed by French psychologists Alfred Binet and Théodore Simon. Those were tests specifically developed to identify mentally deficient children by estimating their mental age, defined as that of the chronologically uniform group of children the average test score of whom a child matched. Thus, if eight-year-olds matched the test scores of ten-year-olds, they had a mental age of ten; if they matched the test scores of five-year-olds, they had a mental age of five. All those children who were thus found to have a mental age that was lower than their chronological age were considered to be mentally deficient. Goddard applied the tests to children in the Vineland Training School for the Feeble-Minded and found them to be very accurate, as they matched the staff's experience with them. Whereas there were children up to 15 years old, none was found to have a mental age greater than 12. Goddard even developed a scale to categorize the "feebleminded": those with a mental age between one and two were classified as "idiots"; those with a mental age between three and seven were classified as "imbeciles"; and those with a mental age between eight and twelve as "morons" (from the Greek word "μωρός", meaning "dull").[31]

In 1909, Goddard was actively looking for funds to support his research. He intended to hire assistants—field workers—who would collect data on heredity by visiting the families of the children in the Vineland Training School. In March of the same year, he received a letter from Davenport, who was asking for "heredity data concerning feeble mindedness". Goddard replied that the available data were unreliable, given that parents would say whatever they would need to say for their child to be accepted to the Vineland Training School. He thus shared with Davenport his intention to hire assistants to collect more reliable data. Davenport expressed his enthusiasm about this prospect and actually visited the Vineland Training School to meet Goddard in April 1909. The collaboration that thus began was mutually beneficial. Whereas Davenport was looking for data, Goddard needed scientific information and especially a clear account of heredity. Davenport recommended that Goddard read Punnett's book *Mendelism* (see Chapter 6). This was the work from which Goddard acquired his understanding of heredity, which was one that clearly distinguished between biological and social inheritance. Goddard found Punnett's book both fascinating and informative and became convinced that Mendelism was applicable to human mental traits.[32]

Davenport had already assumed that "feeblemindedness" was a recessive trait for which Mendelian ratios might be found. Therefore, he wrote to Goddard on July 7, 1909, that it was particularly desirable to "collect facts supporting the conclusion that the offspring of two imbecile parents are all imbecile".[33] As historian Leila Zenderland has pointed out, the hypothesis that "feeblemindedness" was a unit character guided all their research, even though both Davenport and Goddard had claimed that they were not sure if this was really the case.[34] Most interestingly, we see here Davenport taking a rather unscientific way of approaching the problem. Scientists collect data and then analyze those in order to draw conclusions, being open-minded with respect to what these conclusions might be. If a hypothesis was tested, it might be rejected; if a research question had been asked, scientists might find answers in the available data. But Davenport did none of this. Instead, as his statement indicated, he seems to have reached a conclusion already—"feeblemindedness" is inherited as a recessive trait—and then look for data to support this conclusion. The danger inherent in this attitude is that it may be tempting to cherry-pick (even unconsciously) those data that fit with the conclusion already reached, while disregarding those that do not.

Soon after visiting Goddard, Davenport asked him to write a preliminary report summarizing the available knowledge on "feeblemindedness", which was presented in a meeting of December 1909. By that time, Goddard had become secretary of the sub-committee on "feeblemindedness" of the American Breeders' Association.[35] In a paper published a few months later in the journal of that association, Goddard set out to explain the heredity of "feeblemindedness". He presented various family trees that he considered as typical of the about 80 completed by that time, with several of them (Charts I, VIII, X, XI, XIII, and XV) clearly showing various cases where the mating between two "feebleminded" persons had resulted in all of their offspring being "feebleminded". Goddard considered this as the norm, with statements like this: "Chart I shows the maternal grandparents feeble-minded, and they have as usual only feeble-minded offspring—two girls".[36] Those were the results that Davenport had been looking for, showing that all the offspring of two mentally defective people would also be mentally defective. There were other implicit statements about "feeblemindedness" being a recessive trait. Chart VI showed what Goddard described as "a marked instance of the defect skipping a generation".[37] A "feebleminded" woman had several children, none of which was "defective". However, several of her grandchildren were "feebleminded". This could be a typical Mendelian case in which a recessive trait "disappears" in one generation and "reappears" in the next one. But Goddard ended the paper in a cautious tone, noting that this was work in progress and that when it would be completed, it might give "enough data to deduce something of importance concerning human heredity".[38] We may thus think of the year 1909 as marking the beginning of "Mendelian" eugenics. But this was not the only important development.

Historian Peter Bowler has argued that eugenics demanded theories that stressed the role of heredity over the environment, and it is no coincidence that the concept of "pure heredity" was clarified around the same time.[39] Indeed, it was in 1909 that prominent geneticists such as William Bateson and Wilhelm Johannsen clarified this concept. In his 1909 book *Mendel's Principle of Heredity*, Bateson clarified what it meant to be "purebred":

> We at length understand the physiological meaning of "pure-bred" and "cross-bred". We know that these ideas must be applied to the several characters of the animal or the plant, rather than to the individual as a whole. For the individual to be altogether pure-bred it must be homozygous in all respects. In current parlance, dogs, for example, derived from a cross a few generations back have been spoken of as 1/8 Bulldog, or 1/32 Pointer blood, and so forth. Such expressions are quite uncritical, for they neglect the fact that the characters may be transmitted separately, and that an animal may have only 1/32 of the "blood "of some progenitor, and yet be pure in one or more of his traits.[40]

So, an individual could be purebred in different ways; not only "altogether pure-bred" by being homozygous for all characters, but also partially purebred by being homozygous for some traits only. In either case, being "pure" required homozygosity. Also in 1909, Wilhelm Johannsen gave a definition of racial purity in his book *Elemente der exakten Erblichkeitslehre* (*Elements of the Exact Theory of Heredity*). He used the word "rassenrein", meaning "racially pure" or "pure-bred", and citing Bateson for his terms "homozygote" and "heterozygote" he wrote:

> A homozygous being is thus produced from the union of gametes, which brought along the same genes, and is therefore to be called pure-bred or racially pure. A heterozygous being is produced from gametes which were not identical in terms of genes.[41]

We thus see what seems like an important conceptual shift occurring in 1909. Until that time, Mendelian genetics was a model that had been used quite productively in research with domesticated animals and plants. As we saw in Chapter 6, mostly with the work of Bateson and his collaborators, Mendelian genetics was fruitful enough to open new areas of inquiry, as well as flexible enough to accommodate exceptions. But it was used for research only; it was a way to do genetics research with model systems. It had limited application in the study of human heredity. As Bateson also wrote in 1909: "Of Mendelian inheritance of normal characteristics in man there is as yet but little evidence". He added that the only clear case was eye color and that the lack of evidence was probably due to the difficulties in studying heredity in humans, such as having a small number of offspring and large generation time compared to animals and plants.[42] But in 1910, along with the establishment of the Eugenics Record Office, the situation changed. Within a few years, what had been a model for research was transformed into the main explanatory scheme for the heredity of human mental traits and behavior. What had been a useful tool for scientific exploration became a means for ideological confirmation. Davenport and Goddard had a key role in this shift.

Why did this happen? Galton's initial approach to eugenics had nothing to do with unit characters, and as we saw it was developed well before Mendel's work was "re-discovered". At the beginning of the 20th century, Davenport was regarded as a prominent biometrician, and Pearson was the one to continue to provide a biometrician justification for eugenics. Even Lamarckian mechanisms had been invoked to justify eugenics. However, in the end, and after considerable debate, it was Bateson's Mendelism, and his notion of unit characters, which was widely accepted as the most effective model.[43] This was in part due to the influence of Davenport; however, it was Goddard who made "Mendelian eugenics" more widely known. Like Galton, who had studied twins reared apart, Goddard was looking for a similar natural experiment. He found this in the pseudonymous Kallikak family.[44]

A NATURAL EXPERIMENT IN THE STUDY OF "FEEBLEMINDEDNESS"

There were around 400 students in the Vineland institution, and by studying their ancestry Goddard and his associates were able to form a detailed pedigree for one of them: Deborah Kallikak (that was a fictitious name). The findings and the conclusions from this study were presented in Goddard's 1912 book *The Kallikak Family: A Study in the Heredity of Feeble-Mindedness*. The Kallikaks were a family with two lines, a "good" and a "bad" one (hence the family name, according to historian

Leila Zenderland, from the Greek words "kallos" for beauty and "kakos" for bad).[45] The key figure in this story was Martin Kallikak, who started these two family lines. The "bad" family line started in 1776, when Martin joined the Revolutionary Army. In a tavern where soldiers used to hang out, he met a "feebleminded" (anonymous) girl and fathered her "feebleminded" son. The girl named the boy after his father, and so he became Martin Kallikak Jr, who was the great-great-grandfather of Deborah. Among his 480 descendants, Goddard found "conclusive proof" that 143 of them were feebleminded, whereas only 46 were found to be normal, with the rest being unknown or doubtful. Among these 480 descendants, Goddard added, 36 were illegitimate, 33 were "sexually immoral" (mostly prostitutes), 24 were alcoholics, 8 kept houses of ill fame, 3 were epileptics, 3 were criminal, whereas 82 died in infancy. In 1779, after leaving the Revolutionary Army, Martin married "a young woman of a good Quaker family".[46] It is through that union that the "good" family line emerged, which according to Goddard produced "descendants of radically different character": all were "normal people". Among 496 of them, only 3 had been found to be "somewhat degenerate" but not defective: 2 were alcoholics and the 3rd was "sexually loose".[47]

Goddard described the Kallikak family as presenting "a natural experiment in heredity".[48] He believed that this was an experiment in which the environment had been kept constant naturally. As he explained, the members of the two lineages lived out their lives "in practically the same region and in the same environment". In one case, they were so close that a man from the "bad" line had been employed by a family on the "good" line, without anyone suspecting a relationship in spite of having the same name. Goddard concluded that "The biologist could hardly plan and carry out a more rigid experiment or one from which the conclusions would follow more inevitably".[49] The conclusions were thus indisputable for Goddard. "Feeblemindedness" was a condition that could not be changed by the environment, such as by educating those people. It was inherited like other physical features: "no amount of education or good environment can change a feebleminded individual into a normal one, any more than it can change a redhaired stock into a black-haired stock", Goddard wrote. For him the "enormous proportion of feeble-minded individuals" in the "bad" line and their total absence in the "good" line of the Kallikak family was a "striking fact" that was "conclusive on this point".[50] These two lines were represented with two charts in the book. Chart II (Figure 8.4) represented the immediate generations, with a focus on the "bad" line that started with Martin Jr, the "feebleminded" son of Martin and the "feebleminded" woman. Whereas chart I (Figure 8.4) represented the descendants in several subsequent generations, reaching people living at the time that Goddard wrote the book. On the "bad" side, there was Deborah Kallikak, whereas on the "good" side, there was "the son of a prominent and wealthy citizen of the same family name, now resident of another State".[51]

Goddard's book contains a detailed discussion of Mendel's work, his "law" and its application to humans, in a section titled "The Mendelian Law". After describing results with "tall" and "dwarf" peas, and defining the "dominant" and "recessive" characters, Goddard noted: "The recessive factor is now generally considered to be due to the absence of something which, if present, would give the dominant factor", which brings to mind Bateson's presence/absence theory (see Chapter 6). After

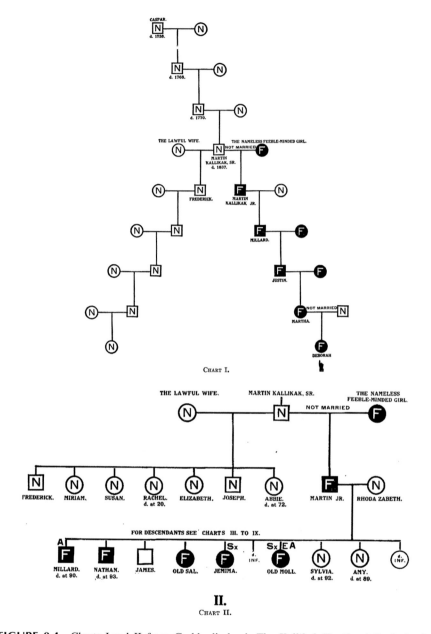

FIGURE 8.4 Charts I and II from Goddard's book *The Kallikak Family: A Study in the Heredity of Feeble-Mindedness*, representing the "good" and the "bad" sides of the Kallikak family lines. Chart I shows the "good" and "bad" lines of the Kallikak family, each of which was traced through the line of the eldest son down to a Goddard's contemporary. Chart II shows the children of Martin Kallikak by his wife, as well as those by his son, also called Martin, from his affair with the anonymous "feebleminded" girl. Black squares and circles with a white "F" indicated feebleminded individuals, whereas "N" indicated normal persons. From Goddard (1911), pp. 36–37. Public Domain.

mentioning that this law applied to various characters in animals and plants, as well as in humans, Goddard noted:

> We do not know that feeble-mindedness is a "unit character". Indeed, there are many reasons for thinking that it cannot be. But assuming for the sake of simplifying our illustration that it is a "unit character", then we have something like the following conditions. If two feeble-minded people marry, then we have the same unit character in both, and all of the offspring will be feeble-minded; and if these offspring select feebleminded mates, then the same thing will continue. But what will happen if a feeble-minded person takes a normal mate? If feeble-mindedness is recessive (due to the absence of something that would make for normality), we would expect in the first generation from such a union all normal children, and if these children marry persons like themselves, i.e. the offspring of one normal and one defective parent, then the offspring would be normal and defective in the ratio of three to one. Of the normal children, one third would breed true and we would have a normal line of descent.[52]

So despite admitting not only that there was not sufficient evidence that "feeble-mindedness" was a recessive character but also that there were reasons to think that this was not really the case, Goddard went on to apply classic Mendelian rules in a crossing. He then wondered whether sterilizing a "feebleminded" person would make any big difference, as only one in four of their grandchildren were expected to be defective. However, he noted that this was an average and that it was possible for a "feebleminded" person to have only one child that would be "feebleminded" too.

It is interesting that whenever Goddard encountered exceptions to Mendelian rules, he dismissed them. For instance, at some point in the book he described the case of Martin Jr.'s oldest son, Millard, also an ancestor of direct descent of Deborah. Millar had married Althea Haight and had 15 children. As both of them were "feebleminded", "according to Mendelian expectations", Goddard wrote, all of their children should have been "feebleminded" too. This expectation was not confirmed by the facts, as there was an exception. Their fourth child, a girl, was taken into another family and grew up being a "normal woman". Her marriage with a "normal man" resulted in descendants who "were normal and above average intelligence".[53] Goddard did not worry much about this exception, though. But if "feeblemindedness" was really due to a recessive factor, it is basic Mendelian genetics that two persons homozygous for the recessive allele cannot have a child with the dominant phenotype. Goddard did not tell us whether this exception was due to the environment in which the girl was brought up or a mutation.

In his conclusion, Goddard expressed his certainty that "feeblemindedness" was a hereditary character: "Feeble-mindedness is hereditary and transmitted as surely as any other character. We cannot successfully cope with these conditions until we recognize feeblemindedness and its hereditary nature, recognize it early, and take care of it". How could they take care of it? Sterilization could be a solution. Yet, Goddard also noted that this was just a "makeshift and temporary" solution, because a lot was still to be learned about "the laws of human inheritance" (!)[54] Whether he was cautious or just inconsistent, we cannot know. But what we know according to historian Leila Zenderland is what Goddard did not write in this book. The book was free from the nativist and racist statements found in the writings of other eugenicists of the time; there was nothing about racial miscegenation, threat from immigration, Anglo-Saxon superiority, or white supremacy. In fact, the "feebleminded" Kallikaks themselves were white, Anglo-Saxon Protestants who had been living in America since the time of the Revolution. The word "eugenics" does not even appear in the book, whereas he made a quick reference to "euthenists" who believed that the environment was the only important factor.[55]

Perhaps because Goddard did indeed consider the data presented in the *Kallikak Family* book as being inconclusive, he continued his studies, the results of which were presented in a huge book titled *Feeble-Mindedness: Its Causes and Consequences*, published in 1914. The title of Chapter VII explicitly asked the question: "Is Feeble-Mindedness a Unit Character?" Goddard answered positively: "That feeble-mindedness is hereditary is abundantly demonstrated from the case histories presented". However, in the very next sentence, he made a statement that could be considered self-contradicting: "Feeble-mindedness is most naturally considered as a lack of intelligence; from this standpoint we would expect that intelligence is dominant, but it is hard for psychologists to think of intelligence as a unit". He went on to explain that a person could be intelligent along some lines but not along some others. So, one might wonder: if intelligence was not a unit, why the lack thereof was? Goddard suggested that based on the available facts, one could think "that feeble-mindedness is a unit character, and due either to the presence of something which acts as an inhibitor, or due to the absence of some stimulus which sends the normal brain on to further development". Considering the latter option as easier to conceive, Goddard added that in such a case one might then think of intelligence as a dominant unit factor.[56]

In the next chapter, he attempted to decide definitively between the two options by analyzing statistically the results of various matings. What had been observed was very close to what would have been expected based on Mendelian genetics. For an expected number of 704 "feebleminded", there were actually 708; for an expected number of 352 "normal", there were actually 348. Goddard's conclusion was that as the results could not be explained in any other way, the only conclusion was that "normalmindedness is, or at least behaves like, a unit character; is dominant and is transmitted in accordance with the Mendelian law of inheritance". Interestingly, he added: "The writer confesses to being one of those psychologists who find it hard to accept the idea that the intelligence even acts like a unit character. But there seems to be no way to escape the conclusion from these figures".[57]

Not surprisingly, Davenport and others took Goddard's conclusions very seriously.

THE WAR AGAINST THE "FEEBLEMINDED"

In 1911 Davenport published a book titled *Heredity in Relation to Eugenics*, which he dedicated to Mrs. E. H. Harriman "in recognition of the generous assistance she has given to research in eugenics". In that book, Davenport explained the nature, importance, and aims of eugenics, as well as its methods, and then presented his findings on the inheritance of family traits such as eye color, hair color, hair form, skin color, stature, body weight, musical ability, ability in artistical or literary composition, mechanical skill, calculating ability, memory, temperament, handwriting, general bodily energy and strength, and general mental ability. There were also accounts of epilepsy, insanity, pauperism, narcotism, criminality, and then of many kinds of human disease. In Chapter 2, Davenport presented the method of eugenics as being based on "the so-called Mendelian analysis of heredity".[58] In Chapter 3, he turned to the inheritance of family traits, and eye color was the first to be considered. It should be noted that at the beginning of the book, there is a plate illustrating the variation of eye color, starting from the "albino" eye that had no pigment and going through the blue, green, hazel, or gray to the brown and the black eye. Davenport analyzed available data by several researchers, including work done by himself and his wife Gertrude.[59] His conclusion, as before, was that blue eyes were inherited in a Mendelian recessive manner: "Brown-eyed children can be secured from blue-eyed stock by mating with pure brown-eyed stock. We have heard of two blue-eyed parents regretting that they had no brown-eyed children. They wished for the impossible".[60] However, one of the studies that Davenport cited had made a clear statement about such a possibility. As the authors had written: "Blue x blue as a rule gives only blue. However the rule is not without exceptions. In one case a man with brown eyes was born of blue-eyed parents".[61] Davenport nevertheless dismissed this result by writing that this "discordant case" was "probably due to an error".[62] When a result does not fit your conclusion, you can easily dismiss it, can't you?

In the same book, Davenport also considered "feeblemindedness" citing the previous work of Goddard. Davenport described the laws of inheritance of mental traits in humans:

> *Two mentally defective parents will produce only mentally defective offspring.* This is the first law of inheritance of mental ability. . . . The second law of heredity of mentality is that, aside from "mongolians", probably no imbecile is born except of parents who, if not mentally defective themselves, both carry mental defect in their germ plasm.[63]

Davenport thus arrived at a conclusion: "In view of the certainty that all of the children of two feeble-minded parents will be defective how great is the folly, yes, the crime, of letting two such persons marry".[64] The next year a book was published, titled *Heredity and Eugenics*, which contained essays based on lectures given by William Castle, John Coulter, Charles Davenport, Edward East, and William Tower during the summer of 1911 at the University of Chicago. In one of his essays titled "The Inheritance of Physical and Mental Traits of Man and Their Application to Eugenics", Davenport referred to the work of Goddard and concluded:

> From the studies of Dr. Goddard and others, it appears that when both parents are feeble-minded all of the children will be so likewise; this conclusion has been tested again and again. . . . But if *one* of the parents be normal and of normal ancestry, all of the children may be normal . . . whereas, if the normal person have defective germ cells, half of his progeny by a feeble-minded woman will be defective.[65]

Simple Mendelian genetics, isn't it?

It is worth noting though that Davenport was not a proponent of sterilization. Already in 1911, he had actually questioned whether it was the best method.

There is no question that if every feeble-minded, epileptic, insane, or criminalistic person now in the United States were operated on this year there would be an enormous reduction of the population of our institutions 25 or 30 years hence; but is it certain that such asexualization or sterilization is, on the whole, the best treatment? Is there any other method which will interfere less with natural conditions and bring about the same or perhaps better results?[66]

He even suggested that there was at the time a lack of sufficient understanding for such measures to be taken. First, Davenport noted that the sterilization legislation was not in agreement with what was known about heredity. "While a feeble-minded person lacks, *ipso facto*, the determiner for normal development, in his germ cells, still we do not know that his children will be defective". Second, "the laws against the marriage of the feebleminded are unscientific because they attempt no definition of the class". The condition was not clearly distinct from normality as clearly as, for instance, polydactylism was. Third, he wondered if there were good grounds for denying marriages among those people. "Is it desirable to encourage non-legal and irregular unions to sustain a law passed without inquiry and based on no certain knowledge?" For Davenport, a better alternative to sterilization was the segregation of the "feebleminded" below a certain grade during the reproductive period, in spite of being more expensive.[67] Laughlin, Davenport's close collaborator did agree that segregation, for life or during the reproductive period, should be "the principal agent used by society in cutting off its supply of defectives". He suggested that sterilization should be advocated "only as supplementary to the segregation feature of the program, which is equally effective eugenically, and more effective socially".[68]

The first sterilization laws in the United States were introduced in Indiana in 1907 and in Washington and California in 1909. In March 1907, Dr. Harry Sharp, the medical superintendent at the Indiana Reformatory in Jeffersonville, who had already performed vasectomies on prisoners since 1899, and William H. Whittaker, the general superintendent of the Indiana Reformatory, introduced an act to "prevent the procreation of confirmed criminals, idiots, imbeciles, and rapists".[69] Sharp believed that the transmission of mental and physical defects would thus stop, and by 1909 he reported that he had sterilized 456 men. However, in the same year Indiana's governor, Thomas R. Marshall, ordered a moratorium on sterilizations being unsure of the statute's constitutionality. It was only in 1919 that Governor James Goodrich decided to test the law's constitutionality. The Indiana Supreme Court decided that the sterilization law violated not only the state constitution but also the U.S. Constitution, specifically the Fourteenth Amendment. This limited the number of sterilizations in the state. In California, things were different. Signed into law by the governor in April 1909, although it initially targeted the "feebleminded", by 1913 it was broadened to apply to inmates and patients. This modification also legally protected the officials participating in sterilization procedures, regardless of patient consent. Another modification in 1917 made sterilization a precondition for institutional release. By 1921, 2,248 sterilizations had been performed in California, which represented more than 80% of all cases in the United States.[70] California was the leading state in the United States with respect to sterilizations, but the total number performed was relatively small, compared to what happened later.

To advance their cause, eugenicists needed to make it acceptable and convince more people. During the second International Congress for Eugenics that took place at the American Museum of Natural History in New York in 1921, various exhibits were used to promote eugenics and gain support for the movement. A key feature of that congress was the "certificate awarded for exhibits and services of merit to the Exhibition", which was found on page 15 of the conference book published in 1923 (Figure 8.5). Signed by the president of the congress palaeontologist Henry Fairfield Osborn, and the chairman of the committee on exhibits, Harry Laughlin, the certificate expressed the acknowledgment of the organizers to Mrs. Harriman for her financial support. Its main feature was the tree of eugenics on its upper part. As it was stated, eugenics, like a tree, drew its materials from various disciplines. As historian Joe Cain has suggested, the intention behind the metaphor of a tree with many roots and branches was to represent eugenics as a multidisciplinary subject and thus to encourage participation from many types of experts. The key message? "Eugenics is the self direction of human evolution". Humans could henceforth direct their own evolution wherever they wanted.

Given the central place of Mendelian genetics in eugenic arguments, it was necessary for eugenicists to clearly explain Mendelian genetics. They did this with a variety of means. An interesting one was a mechanical display called Mendel's theatre, which we see in Figure 8.6 operated in 1926 by Leon Fradley Whitney, a prominent member of the American Eugenics Society, the Eugenics Research Association, and the American Genetics Association. It was used to educate potential field workers who would help direct future progress toward human betterment. To make visible what was invisible, the eugenicists relied heavily on performance and theatricality. This display was a large white box in the shape of a theatre stage, and it was filled with a rolling scroll that displayed three acts in turn: A, B, and C. Those acts displayed pictures and dolls of the Mendelian inheritance of human hair and skin color.[71] In the scene shown in Figure 8.6, we see Act C and the display of a typical case of Mendelian inheritance. Two parents with dark hair had four children: three with dark hair and one with lighter hair. This is the typical Mendelian 3:1 ratio, which however might never appear in a family with four offspring due to chance. This is a common misunderstanding among students, as the Punnett squares only display the probabilities for a particular outcome, and not the actual proportions. It is only when the number of offspring is quite large that we might observe proportions that correspond closely to the expected ones.

Eugenicists also used particular exhibits in exhibitions and fairs in order to illustrate the laws of Mendelian inheritance. A typical example was an array of stuffed black and white guinea pigs arranged on a vertical board in a way that showed the inheritance of coat color. On the left side was the parental generation, whereas on the right side was its progeny in numbers that represented the Mendelian ratios.[72] Understanding the main principles of Mendelian genetics, and the various "Mendelian Possibilities", as the outcomes of the various possible crosses were described on the array with the guinea pigs on the left part of Figure 8.7, was obviously considered essential for making people understand how heredity worked and how the various "bad" recessive traits were propagated by those who had them. As shown in that array, a cross between a white and a black parent gave only black offspring. This

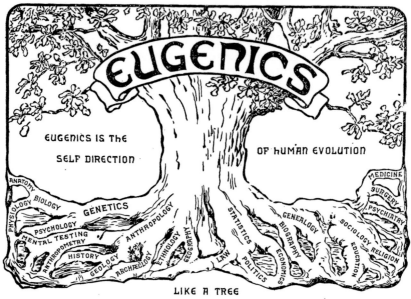

EUGENICS

EUGENICS IS THE SELF DIRECTION OF HUMAN EVOLUTION

LIKE A TREE
EUGENICS DRAWS ITS MATERIALS FROM MANY SOURCES AND ORGANIZES
THEM INTO AN HARMONIOUS ENTITY.

*The Second International Congress of Eugenics,
devoted to researches in all fields of science and
practice which bear upon the improvement of racial
qualities in man, convey this expression of appre-
ciation of the generous gift of*

Mrs E. H. Harriman

*of New York, which made possible the exhibition
of eugenical materials which were assembled and
displayed, in connection with the Congress, at the
American Museum of Natural History.*

New York, September 1921

President of the Congress.   Chairman of the Committee on Exhibits

FIGURE 8.5   The certificate awarded for exhibits and services of merit to the exhibition at the second International Congress for Eugenics. From Laughlin (1923), page 15, figure 3. Courtesy of Cold Spring Harbor Laboratory Archives.

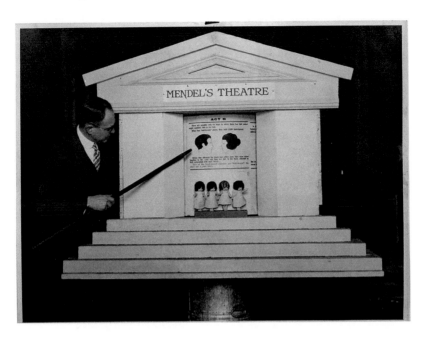

FIGURE 8.6 "Mendel's Theatre" showing the inheritance of hair color, demonstrated by eugenicist Leon Fradley Whitney (1894–1973) in 1926. Courtesy of the American Philosophical Society.

meant, according to the Mendelian genetics interpretation, that black was dominant and white was recessive. As a result, the cross between "pure white" parents would always yield offspring of the same kind. But the real problem was that the cross between two black heterozygotes could yield white offspring—which transferred to the case of "feeblemindedness" meant that even two rather normal individuals could have a "feebleminded" child.

This understanding of the Mendelian genetics of guinea pig color was essential for convincing people about the value of eugenics. As a chart put it, "Unfit human traits such as feeblemindedness, epilepsy, criminality, insanity, alcoholism, pauperism, and many other run in families, and are inherited in exactly the same way as color in guinea pigs" (Figure 8.8). Here is the real problem with the direct application of Mendelian genetics from a single trait studied in the lab to complex human traits and behaviors. If people had read the writings of Morgan presented in the previous chapter, they would have questioned this idea. But for people without such knowledge, it was easy to accept the idea that complex human traits are inherited in exactly the same way as the color in guinea pigs when this was presented as a scientific fact. Based on this, the chart arrived at the conclusion that if all marriages had been eugenic ones, most of these unfit traits would be "bred out" in three generations. On the right side of the chart comes the misleading message: "You can improve your education, and even change your environment; but what you really are was all settled when your parents were born" (Figure 8.8). In other words, whatever you do, there is

FIGURE 8.7   Eugenic and Health Exhibit, Kansas Free Fair, 1929. On the leftmost part of the photo, there is an array of stuffed black and white guinea pigs arranged on a vertical board in a way that shows the inheritance of coat color. Courtesy of the American Philosophical Society.

little hope. Better children can only come from selected parents, and this was the aim of eugenics, according to the chart.

This argument for eugenics against the "feebleminded", and others, in order to improve the genetic basis of the population was not the only one. Perhaps on its own was not even convincing. This is why it was often accompanied by an economic argument. Simply put, this argument was the following: there were many "feeble-minded" in society who were unable to support themselves and who often caused problems. For one or both of these reasons, these people had to be institutionalized so that they would be taken care of and controlled. But this would cost money, and who would pay that cost? Taxpayers. Therefore, rather than letting these "feebleminded" people procreate their kind, the argument went, and then tax the rest of the population in order to support their miserable lives, it would be preferable to not let them be born at all. This would save society from trouble and wasting money. As Davenport himself put in a 1911 report of the committee of eugenics:

> If only one-half of 1 per cent of the $30,000,000 annually spent on hospitals, $20,000,000 on insane asylums, $20,000,000 for almshouses, $13,000,000 on prisons, and $5,000,000 on the feeble-minded, deaf and blind were spent on the study of the bad germ-plasm that makes necessary the annual expenditure of nearly $100,000,000 in the care of its produce we might hope to learn just how it is being reproduced and the best way to diminish, its

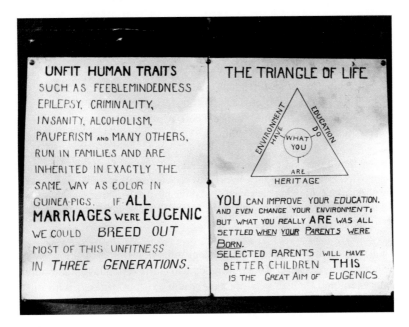

FIGURE 8.8 Chart used at the Kansas Free Fair describing "unfit human traits" and the importance of eugenic marriage. Courtesy of the American Philosophical Society. The chart (misleadingly) informs readers that "Unfit human traits such as feeblemindedness, epilepsy, criminality, insanity, alcoholism, pauperism, and many other run in families, and are inherited in exactly the same way as color in guinea pigs". Courtesy of the American Philosophical Society.

further spread. A new plague that rendered 4 per cent of our population, chiefly at the most productive age, not only incompetent but a burden costing $100,000,000 yearly to support would instantly attract universal attention, and millions would be forthcoming for its study as they have been for the study of cancer. But we have become so used to crime, disease and degeneracy that we take them as necessary evils. That they were, in the world's ignorance, is granted. That they must remain so, is denied.[73]

It was in this way that a strong emphasis on the power of heredity offered a cheaper solution for addressing a societal problem. Any kind of social reform or intervention would be a waste of money if "inferior ability" was hereditary. Therefore, the only solution was to prevent those people of "inferior ability" from propagating their kind.

What was the price to pay in order to improve the "stock" of the nation and save money? Based on various reports, it is estimated that in the decades before World War II, at least 60,000 institutionalized persons were sterilized in the United States. Until 1939, about 30 states had enacted laws that allowed sterilization by the authorities of individuals in state homes for the mentally disabled and state hospitals for the mentally ill who were considered unfit to be parents. Initially, state programs had focused on sterilizing men, but after 1930 mentally disabled women were sterilized as well.[74] A key event in the history of eugenic sterilization began in Virginia in 1924, when state officials convinced an attorney to represent the interests of a young woman named Carrie Buck in order to test the constitutionality of a recently enacted law. Buck had been classified as "feebleminded" and unfit for parenthood and therefore eligible for sterilization. The case became known as *Buck v. Bell*, John Hendren Bell being the superintendent of the Virginia State Colony for Epileptics and Feebleminded where Carrie had been transferred. Eventually, the case reached the Supreme Court of the United States. There was an 8–1 ruling, with the famous statement by Judge Oliver Wendell Holmes:[75]

> It is better for all the world if, instead of waiting to execute degenerate offspring for crime or to let them starve for their imbecility, society can prevent those who are manifestly unfit from continuing their kind. The principle that sustains compulsory vaccination is broad enough to cover cutting the Fallopian tubes. . . . Three generations of imbeciles are enough.[76]

A consequence of this decision was the legitimization of sterilization laws in the United States. Whereas several states already had sterilization laws, their application was inconsistent and their effects practically non-existent, perhaps with the exception of California. After *Buck v. Bell*, many states accordingly updated and put into effect their sterilization legislation.

It is worth noting that Mendelian genetics was invoked via the Kallikak family story to explain "feeblemindedness" for the Buck case in the Virginia court in 1924. In his testimony as one of Judge Aubrey Strode's expert witnesses, noted Virginia eugenicist Dr. Joseph S. DeJarnette, Superintendent of the Western State Hospital at Staunton, Virginia, stated that if both parents were "feebleminded", it was "practically certain" that the children would all be "feebleminded". When he was subsequently asked what would happen if the feebleminded woman was mated with a normal man, his response was: "Then about one-fourth of them would probably be feeble-minded". Another question was about the likelihood, given DeJarnette's experience, for a "feebleminded" woman to mate "with people of her own type, or normal men". DeJarnette's response was "The feebleminded woman will have three children to every one child a college graduate will have. They are easily over-sexed, and it depends on their looks as to how the boys or men will take advantage of them, and it depends on her opportunities". And what was the evidence for this claim? "The Callicac case". DeJarnette referred to "Old man John Callicac" and the familiar story from Goddard's book. After providing a rough description of the two lines, stating that there was no "feebleminded" descendant in the "good" line in contrast to the "bad" line, he justified his reference to this case by stating: "That is a report that was generally published throughout most of the books on heredity".[77]

DeJarnette was also an ardent Mendelian and very proud of the following poem he wrote during the 1920s:

Mendels Law: A Plea For A Better Race Of Men

Oh, why are you men so foolish —
You breeders who breed our men
Let the fools, the weaklings and crazy

Keep breeding and breeding again?
The criminal, deformed, and the misfit,
Dependent, diseased, and the rest —
As we breed the human family
The worst is as good as the best.
Go to the house of some farmer,
Look through his barns and sheds,
Look at his horses and cattle,
Even his hogs are thorough breds;
Then look at his stamp on his children,
Low browed with the monkey jaw,
Ape handed, and silly, and foolish —
Bred true to Mendel's law.
Go to some homes in the village,
Look at the garden beds,
The cabbage, the lettuce and turnips,
Even the beets are thoroughbreds;
Then look at the many children
With hands like the monkey's paw,
Bowlegged, flat headed, and foolish —
Bred true to Mendel's law.
This is the law of Mendel,
And often he makes it plain,
Defectives will breed defectives
And the insane breed insane.
Oh, why do we allow these people
To breed back to the monkey's nest,
To increase our country's burdens
When we should only breed the best?
Oh, you wise men take up the burden,
And make this you loudest creed,
Sterilize the misfits promptly —
All not fit to breed!
Then our race will be strengthened and bettered,
And our men and our women be blest,
Not apish, repulsive and foolish,
For the best will breed the best.[78]

What we have here is a manifesto for eugenics citing "Mendel's law". The argument is not new, of course, and essentially goes back to Galton and the origins of eugenics. If we can successfully breed all kinds of animals and plants with the desired traits, shouldn't we be doing the same for humans? Why do we let "defectives" and "insane" propagate their kind, when "we should only breed the best?" What should we do? "Sterilize the misfits" and let the best "breed the best". But what is interesting is the underlying rule for all this "Mendel's law". "Bad" will only produce "bad"—"Bred true to Mendel's law". In genetics parlance, this meant that defective individuals are "true-breeding", which means that they always pass down to their descendants the same (defective) phenotypic traits—something that happens for instance when individuals are homozygous recessive. However good DeJarnette's understanding of Mendelian genetics was, what is interesting here is that we have a popularized version of the key idea of "Mendelian" eugenics.

You may now reasonably wonder: What did geneticists have to say about all this?

NOTES

1 Galton (1908), p. 308.
2 Kevles (1995), p. 38.
3 www.ucl.ac.uk/news/2020/jun/ucl-denames-buildings-named-after-eugenicists (accessed May 29, 2023).
4 Peterson (2024).
5 Kampourakis (2024).
6 Darwin (1871), p. 168.
7 Darwin (1871), pp. 168–169.
8 Galton (1865), pp. 163–164.
9 Galton (1865), pp. 319–320.
10 Galton (1869), p. 64.
11 Galton (1883), pp. 24–25.
12 Saleeby (1909), p. 172.
13 Galton (1904), p. 1.
14 Galton (1904), p. 3.
15 Galton (1904), pp. 10–11.
16 Galton (1904), pp. 5–6.
17 Galton (1904), pp. 23–24.
18 Radick (2023), p. 206.
19 Galton (1904), pp. 9–10.
20 Kevles (1995), pp. 22–27.
21 Pearson (1906), p. 1.
22 Kevles (1995), pp. 38–40.
23 https://mediacentral.ucl.ac.uk/Play/43712#! (accessed June 12, 2023).
24 Pearson and Moul (1925), p. 6.
25 Pearson and Moul (1925), p. 7.
26 Barker (1989).
27 Gilbert, M. (2009). Churchill and Eugenics. Available at: https://winstonchurchill.org/publications/finest-hour-extras/churchill-and-eugenics-1/ (accessed June 12, 2023); See also Rutherford (2022), Part I.
28 Kevles (1995), Chapter 3.
29 Pauly (2000), p. 222.
30 Davenport (1909), pp. 20–21.
31 Kevles (1995), pp. 77–78.
32 Zenderland (2001), pp. 153–156.
33 Quoted in Zenderland (2001), p. 157.
34 Zenderland (2001), p. 398.
35 Zenderland (2001), pp. 164–165.
36 Goddard (1910), p. 166.
37 Goddard (1910), p. 169.
38 Goddard (1910), p. 178.

39 Bowler (1989), pp. 165–166.
40 Bateson (1909), p. 291.
41 Johannsen (1909), p. 128 (author's translation).
42 Bateson (1909), p. 205.
43 Bowler (1989), pp. 166–167; see also Radick (2023).
44 Zenderland (2001), p. 163.
45 Zenderland (2001), p. 175.
46 Goddard (1912), p. 99.
47 Goddard (1912), pp. 18–19, 29–30.
48 Goddard (1912), p. 116.
49 Goddard (1912), pp. 50–51, 69.
50 Goddard (1912), p. 53.
51 Goddard (1912), p. 33.
52 Goddard (1912), pp. 111–112.
53 Goddard (1912), pp. 22–24.
54 Goddard (1912), pp. 117.
55 Zenderland (2001), pp. 178–179.
56 Goddard (1914), p. 547.
57 Goddard (1914), p. 556.
58 Davenport (1911a), p. 18.
59 Davenport and Davenport (1907), p. 592.
60 Davenport (1911a), p. 31.
61 Holmes and Loomis (1909), p. 52.
62 Davenport (1911a), p. 31.
63 Davenport (1911a), pp. 66–67, emphasis in the original.
64 Davenport (1911a), p. 67.
65 Davenport (1912a, 1912b). pp. 281–282.
66 Davenport (1911a), p. 256.
67 Davenport (1911a), pp. 257–259.
68 Laughlin (1914a), pp. 46–47.
69 Quoted in Stern (2011), p. 98.
70 Stern (2011).
71 Wolff (2009), pp. 61–62.
72 Kevles (1995), p. 62.
73 Davenport (1911b), p. 93.
74 Reilly (2015).
75 Cohen (2017).
76 Statement of Evidence from Hearing on Appeal of Order to Sterilize Carrie Buck: Testimony of Dr. J.S. DeJarnette; 11/18/1924; *Buck v. Bell* (Case File #31681); Appellate Jurisdiction Case Files, 1792–2010; Records of the Supreme Court of the United States, Record Group 267; National Archives Building, Washington, DC., p. 207 (https://catalog.archives.gov/id/45637229; accessed July 13, 2023).
77 Statement of Evidence from Hearing on Appeal of Order to Sterilize Carrie Buck: Testimony of Dr. J.S. DeJarnette; 11/18/1924; *Buck v. Bell* (Case File #31681); Appellate Jurisdiction Case Files, 1792–2010; Records of the Supreme Court of the United States, Record Group 267; National Archives Building, Washington, DC., pp. 74–75 (https://catalog.archives.gov/id/45637229; accessed July 13, 2023).
78 https://encyclopediavirginia.org/wp-content/uploads/2020/11/7552_5599cbee0744352.jpg (Public Domain). This poem was written and published first in a Virginia state government report in 1920 and was found by Paul A. Lombardo in 1981 (personal communication).

9 Geneticists' Attitudes Toward "Mendelian" Eugenics

GENETICISTS ACCEPTING GODDARD'S MENDELIAN RECESSIVE ACCOUNT OF "FEEBLEMINDEDNESS"

In 1989, David Barker at the University of Manchester reviewed the available literature and identified three groups of geneticists[1] with respect to their attitudes toward Mendelian explanations of mental deficiency: (1) those who accepted Goddard's Mendelian recessive account and who argued for tough eugenic measures; (2) those who modified and updated that account in the light of later developments in genetics; and (3) those who criticized it. Table 9.1 presents an overview.[2] We see many big names such as Punnett, Bateson, and East accepting the Mendelian recessive account of "feeblemindedness"; others such as Haldane and Fisher changing their mind across time; and others such as Hogben and Morgan criticizing the idea that "feeblemindedness" is inherited as a recessive Mendelian trait.

As shown in Table 9.1, one consistent supporter of Goddard's theory, and the first among those therein to do so, was the geneticist Edward Murray East (Figure 9.1). Initially trained as a chemist, his first scientific appointment was that of assistant chemist who made chemical analyses of samples of corn. What made East interested in genetics was the interest in the nutritional content of corn, and the proper balance of carbohydrates, fats, and protein. He thus realized that its chemical composition could be improved by breeding. The initial experiments in corn breeding attracted much attention, and when the Connecticut Agricultural Experiment Station, in close collaboration with Yale University, decided to expand their research in plant breeding in that direction, East was recommended for the job and accepted it in 1905. There he conducted many pioneering experiments including the study of Mendelian characters in maize. He also studied the effects of inbreeding and crossbreeding and overall contributed significantly, along with others, to the development of a radically new method of breeding in maize that revolutionized U.S. agriculture. In 1909, East was offered a job at Harvard University, where his research focused on the breeding of tobacco. He stayed there until his death in 1938.[3]

East was a prominent geneticist but also a eugenicist. He had taught with Davenport in that summer course in 1911 at the University of Chicago and contributed essays to the book *Heredity and Eugenics*, published the next year. His essays were about

DOI: 10.1201/9781032449067-11

TABLE 9.1

Geneticists Who Accepted, Modified, or Criticized Goddard's Theory

Geneticists Who Accepted Goddard's Mendelian Recessive Account	Geneticists Who Modified Goddard's Mendelian Recessive Account	Geneticists Who Criticized Goddard's Mendelian Recessive Account
East (1917)	East (1917)	Holmes (1921)
Punnett (1917)	Holmes (1921)	Morgan (1925)
East and Jones (1919)	Fisher (1924)	Hogben (1930)
Gates (1920–1921)		Hogben (1931)
Holmes (1921)	These refinements were mentioned in:	Hogben (1931)
Newman (1921)	Gates (1923 and 1929)	Crew (1931–1932)
Downing (1920)	Crew (1927)	Hogben (1933)
Bateson (1921–1922)	Guyer (1928)	Fisher (1934)
Holmes (1923)	Jennings (1930)	Penrose (1934)
Gates (1923)	Castle (1931)	Haldane (1935)
East (1926)	Huxley (1931)	Haldane (1938)
Crew (1927)		Penrose (1938)
Guyer (1928)		
Haldane (1928)		
Gates (1929)		
Conklin (1929)		
Jennings (1930)		
Huxley (1931)		
Jennings (1931)		
Castle (1931)		
East (1931)		
East (1932)		
Gates (1934)		
Huxley (1937)		
Shull (1938)		
Huxley (1941)		

Source: Based on Barker (1989), p. 362, where the respective references are cited.

heredity and breeding in plants, yet, his second essay titled "The Application of Biological Principles to Plant Breeding" ended with the following conclusion:

> In the field of selection the new ideas are still more economical of time. To the belief that faith and continuous selection toward an ideal would produce any desired result has succeeded the idea that nature alone produces variations and that man's duty is to be alert to grasp their possibilities and to make the most of them. No longer is it believed that many generations of work are necessary to purify a commercial variety of plants from undesirable characters. No longer is there belief that the results of selection are continuous, that it gradually perfects a character. We work for strains homozygous for characters that we know are there, and, by our direct methods we get them without loss of time.[4]

East's message was that we can get pure individuals, homozygous for the desired characters, without spending a lot of time. This was all for plants of course.

FIGURE 9.1   Edward Murray East (1879–1938). Copeland-Bloom Photo Album, Marine Biological Laboratory. Attribution 4.0 International (CC BY 4.0).

In 1917 East wrote an article titled "Hidden Feeble-mindedness". The subtitle is more interesting (looks more like an abstract, though): "One person in fourteen of the American population probably carries the trait in a recessive form, although normal to all appearances—one-fourth of offspring will be feebleminded if mating is made with another carrier". East begun by expressing his concern about the extent of the problem of "feeblemindedness" in the United States, noting that despite some efforts it has not been taken as seriously as it should have been. He credited Goddard as having shown that "feeblemindedness" was transmitted as a Mendelian recessive trait—and did not question that conclusion. In contrast, he took it for granted and went on to explain that "feebleminded" people could emerge in three different ways. The first one was through the mating of two "feebleminded" people, that is homozygous for the trait, in which case all offspring would be "feebleminded" as well. The second case was through the mating of a "feebleminded" with a carrier of the trait (meaning a heterozygous), in which case half of the offspring, on average, would be "feebleminded". Sterilization could have an effect in these two cases. However, there was a third case, which East considered as the most dangerous one: the mating of two carriers of "feeblemindedness", in which case one quarter, on average, of the offspring, would be "feebleminded".

Why was the last case the most dangerous? One reason was that feeblemindedness was "hidden" in the carriers—hence the title of his article—and therefore the mating among those people could not be precluded. However, the most important reason is that those people, the carriers, were many more in the population than the "feeble-minded" themselves. After making calculations, East arrived at the conclusion that if 1 to 200 marriages was between carriers of "feeblemindedness", then 1 in 14 people in the general population must have also been carriers of "feeblemindedness". Given this, in order to eliminate this defective trait, the goal should not be "the comparatively simple one of preventing the multiplication of those so affected. . . . It is rather the almost hopeless task of reducing the birth-rate among transmitters of serious defects". He thus suggested that identifying those people is where efforts should be focused:

> We have assumed that a normal mentality is completely dominant over a defective one. Is this true? Complete dominance is rare among those characters commonly studied by animal and plant geneticists. Is it not likely that the Binet-Simon or other proper tests would show that carriers of mental defects exhibit a lower mentality than pure normals? Would it not be wise to start some investigations along this line?[5]

In other words, even if the "feebleminded" themselves could be identified, this was not enough. Why couldn't be also investigated whether it would be possible to identify the carriers within whom this defective trait was hidden?

Punnett was among the first to agree with East, not only that feeblemindedness is a Mendelian recessive character but also that something more should be done about it. In the Preface of the third edition of his book *Mendelism*, published in 1911, Punnett noted that it was becoming more clearly accepted that the "Eugenic ideal" was mostly defined "by the facts of heredity and variation, and by the laws which govern the transmission of qualities in living things".[6] In a paper presented during the First International Eugenics Congress held at the University of London, July 24 to 30, 1912, Punnett stated:

> The one instance of eugenic importance that could be brought under immediate control is that of feeblemindedness. Speaking generally, the available evidence suggests that it is a case of simple Mendelian inheritance. Occasional exceptions occur, but there is every reason to expect that a policy of strict segregation would rapidly bring about the elimination of this character.[7]

In a 1917 paper titled "Eliminating Feeble-mindedness", Punnett wondered how efficient isolating or sterilizing the individuals that exhibited an undesirable recessive character could be in order to free a population from that. Assuming that "feeble-mindedness" was due to a recessive factor and that mating occurred at random in the population, Punnett calculated how long it would take to diminish its frequency. After presenting his calculations, he concluded that if the proportion of "feeble-minded" in the United States was 3 per 1,000, it would take more than 250 generations, or 8,000 years, to reduce this to 1 in 100,000. This would be very slow and not hopeful. Instead, he suggested another solution:

> Clearly if that most desirable goal of a world rid of the feebleminded is to be reached in a reasonable time some method other than that of the elimination of the feebleminded themselves must eventually be found. The great strength of this defect in the population

lies in its heterozygotic reserves; if the campaign against it is to meet with success it is at these that it must be directed.[8]

Quoting East, Punnett suggested that rather than focusing on "feebleminded" people only, which was insufficient for any change to be brought about any time soon, attention should be directed toward identifying the heterozygote carriers of the conditions. If the Binet-Simon tests, or any other tests of this kind, could identify the heterozygotes—themselves exhibiting lower mental abilities than the "normal" people, then should not something be done? In other words, Punnett argued that eugenic policies should be expanded to more people than the "feebleminded" themselves, in order to accelerate the elimination of "feeblemindedness".

In 1952, on the 50-year anniversary of the publication of Bateson's book *Mendel's Principles of Heredity: A Defence*, which we considered in Chapter 3, the Royal Society asked Punnett's permission to reprint an extract of an article he had written for *The Edinburgh Review* in 1926, soon after Bateson's death. Here is what Punnett had written there:

> The falsity of the doctrine of the equality of men has in modern times been clearly recognized by Gobineau, Galton and many another thinker, but in the absence of any accurate knowledge of heredity their reasoning could make little headway against the prevailing tide of sentimental prejudice. Evolution was preached long before Charles Darwin, though it only became acceptable when he marshalled the facts and forced upon us the logical conclusion. So with that equally fundamental conception of the inborn diversity of man. Running counter as it does to the doctrine of the equality of rights and opportunity which the modern politician, avid of votes, would appear to accept as the basis of the modern State, it is unlikely to meet with general acceptance unless it too is forced upon us by the logic of facts. Not until we had from Bateson the enunciation of discontinuity in variation, and from Mendel the revelation of discontinuity in heredity, could there come into being that irresistible body of facts which must henceforth play so great a part in the destiny of mankind. That the diversity of man is inborn, and no mere accident of surroundings, is no longer a philosophical speculation. It is now certain knowledge, and Bateson insists that it is upon this knowledge that the foundations of a strong and stable State can alone be laid.[9]

Punnett could not have expressed his views any clearer. Differences among humans are inborn, and the doctrine of equality is simply false. Whatever politicians say about equality does not stand the challenge of facts. Since Mendel and Bateson, he argued, it had become clear that human diversity is inherent and does not depend on the environment. It is not only interesting that Punnett held such a view; it is also interesting that the Royal Society had it reprinted in 1952.

What did Bateson have to say about this? Already in 1908, he had written:

> Genetic inquiry aims at providing knowledge that may bring, and I think will bring, certainty into a region of human affairs and concepts which might have been supposed reserved for ages to be the domain of the visionary . . . but so soon as it becomes common knowledge—not a philosophical speculation, but a certainty—that liability to a disease, or the power of resisting its attack, addiction to a particular vice, or to superstition, is due to the presence or absence of a specific ingredient; and finally that

these characteristics are transmitted to the offspring according to definite, predictable rules, then man's views of his own nature, his conceptions of justice, in short his whole outlook on the world, must be profoundly changed. Yet as regards the more tangible of these physical and mental characteristics there can be little doubt that before many years have passed the laws of their transmission will be expressible in simple formulae. . . . If there are societies which refuse to apply the new knowledge, the fault will not lie with Genetics.[10]

Science is clear about heredity. Whether people will listen to it, or not, was their problem.

In his 1909 book *Mendel's Principle of Heredity*, Bateson expressed moderate praise for eugenics. Toward the end of the book, he wrote: "The outcome of genetic research is to show that human society can, if it so please, control its composition more easily than was previously supposed possible". An asterisk at the end of the sentence took the reader to a footnote where Bateson wrote: "Mr F. Galton's long-continued efforts have at length been successful directing public attention in some degree to the overwhelming importance of Eugenics". Bateson noted that whether or how such a form of control should be exercised fell outside the scope of his book. However, he added that because there were already advocates of the application of genetics to human affairs, a few words were appropriate:

Whatever course civilisations like those of Western Europe may be disposed to pursue, there can be little doubt that before long we shall find that communities more fully emancipated from tradition will make a practical application of genetic principles to their own population.[11]

Bateson seemed convinced of the importance of eugenics.

In 1921, an essay by Bateson was published in the *Eugenics Review*, which stemmed from the Galton lecture he had been invited to give by the Eugenics Education Society. Bateson noted that he was never really involved in eugenic activities, nor did he ever become a member of that society. With respect to "feeblemindedness", he noted:

The charge most often brought against the eugenic doctrine is that it aims at perpetuating a rash and subversive interference with habits and manners in pursuit of some cold and calculated purpose. I suppose that that is what some people mean by eugenics. Foolish legislation passed or promoted in certain American States gives colour to such opinions. I have heard also of busybodies who, in the name of eugenics, have made some irresolute young people gratuitously miserable. That crude interpretation is, so far as I see, based neither on scientific fact nor on common sense. Everyone who has studied these problems at all would advise the State to put such control on the feeble-minded members of the population as to prevent their propagation. They are examples of a peculiar physiological condition, not very difficult to recognise, and when they interbreed, as at present they frequently do, they have no normal children but infallibly add to the asylum and institute population. As to the propriety and I may add the humanity of exercising control over these persons we are all agreed, but I know no warrant for direct legislative interference beyond that obvious and altogether special case.[12]

Bateson was an ardent Mendelian but not really an ardent eugenicist. Yet he was not against policies that would put a limit to the procreation of the "feebleminded".

Another interesting figure is Ronald Aylmer Fisher (Figure 9.2). He was an outstanding scholar with important contributions both to statistics and to genetics. On the statistics side, he pioneered the statistical concept of likelihood, extended the use of Student's t-test, developed the theory of significance testing, provided the correct interpretation for the use of Pearson's chi-square goodness-of-fit test, developed an exact test that that now bears his name, and coined the concepts of "variance" and "analysis of variance". On the genetics side, a key contribution was his 1918 paper entitled "The Correlation Between Relatives on the Supposition of Mendelian Inheritance", which showed how Mendelian genetics could explain patterns of correlations in quantitatively varying traits such as height, under the assumption that many different genetic factors contribute to such traits.[13] Fisher is another scholar, along with Galton and Pearson, who was celebrated by UCL, having worked there between 1933 and 1943. He was actually the successor of Pearson as "Francis Galton Professor of National Eugenics", also becoming director of the Francis Galton Laboratory of National Eugenics. In 2020 the UCL Department of Genetics, Evolution and Environment took the decision to de-name the RA Fisher Centre for Computational Biology, to reflect their "concerns over Fisher's support for eugenics

FIGURE 9.2　Ronald Aylmer Fisher (1890–1962). Courtesy of University of Adelaide Library, Rare Books and Manuscripts.

long after it had become an outmoded concept".[14] A window-portrait of Fisher was also removed during the same year from the College Hall of the Gonville and Caius College at the University of Cambridge.[15]

In 1924 Fisher criticized as misleading Punnett's 1917 paper about what it would take to diminish the frequency of "feeblemindedness" in the population. He began by suggesting that the question should be reformulated. What mattered more, he argued, was "the effect of such selection in our immediate posterity . . . the effect which one individual might live to experience". Maintaining Punnett's assumptions, he concluded that in a single generation "the load or public expenditure and personal misery caused by feeblemindedness" would be reduced by 17%. Fisher thus noted that even with Punnett's figures, "it is within our power to leave our children a substantial eugenic legacy. This legacy would be permanent" even if the next generation decided not to diminish the proportion of "feebleminded" people even further.[16] The question for Fisher, with respect to the effectiveness of eugenic measures, was not whether or when "feeblemindedness" would be eliminated. Rather, the question for him was whether a significant effect could be brought about during their lifetime, and his answer was that, even on Punnett's assumptions, this was the case.

Then Fisher turned to Punnett's two assumptions. He acknowledged that there was a considerable body of evidence about "feeblemindedness" being due to a single Mendelian recessive factor. However, he noted that there was an even more substantial body of evidence that the differences among "normal" people or among "feebleminded" people could be as high in degree as those between a "normal" and a "feebleminded" person. Fisher thus concluded that whereas it was possible to think of "feeblemindedness" as due to a Mendelian recessive factor, it was not possible to claim that all "feebleminded" cases were genetically alike. He assumed that there were probably several different Mendelian factors that were responsible. Based on this, Fisher concluded:

> Consequently, in the case of so variable a characteristic as feebleness of mind it would be extremely rash to assume that only one main factor is present, and entirely contrary to the evidence to ignore the contributions of less important factors.[17]

Punnett's second assumption was, according to Fisher, open to more obvious criticism. Fisher argued that mating depended largely on social class, and that within each class there was a tendency for people to mate with other people like them. He gave the example that when it came to human height, the resemblance of husband and wife was nearly as great as that between an uncle and his niece. Therefore, it was more likely for "feebleminded", who gravitated toward the lower social classes, to mate with one another, than with people from another class. Under this assumption, Fisher calculated that the effect of segregation would be a reduction of the incidence of "feeblemindedness" by 36% in one generation and noted: "This is perhaps about what might be expected from an effective policy of segregation, and it is of a magnitude which no one with a care for his country's future can afford to ignore".[18] So we see here that even though Fisher questioned Punnett's estimation, he only did so because it drew wrong conclusions for eugenics. To be clear: Punnett's 1917 paper was an argument for the expansion of eugenics measures and Fisher's criticism of

Punnett's paper was only meant to suggest that the measures already considered, such as the sterilization of the feebleminded, could be quite effective. By the way, it is worth mentioning that Punnett was affiliated with the Cambridge University Eugenics Society that Fisher helped found in 1911.[19]

While this was happening, Goddard's theory fell under fierce attack.

GODDARD'S THEORY UNDER ATTACK

By the early 1920s, the findings of the intelligence tests had become controversial, attracting all kinds of critiques. For example, journalist Walter Lippmann published a series of articles in the journal *New Republic*, in which he questioned the reliability of those tests. In one of these, Lippmann cited the criticisms of psychologist James McKeen Cattell on the Kallikak study. Even though Cattell was a eugenicist and had praised the contribution and the conclusions of Goddard's study, he had pointed out that the evidence for the heredity of "feeblemindedness" would have been more powerful and more interesting from a scientific point of view, if Martin Kallikak's wife had been "feebleminded" and the tavern girl had been a healthy, normal person. In such a case, it would have been possible to argue with confidence that this was a case of biological rather than social heredity.[20] The point made was that if the descendants of Martin Kallikak and a "feebleminded" wife from a noble family were also "feebleminded", notwithstanding the good environment in which they had been brought up, then it would be possible to argue for the hereditary nature of the condition, as it could not have been due to environmental causes.

Among all critiques, perhaps the most devastating one came from the neurologist Abraham Myerson (Figure 9.3) in his 1925 book *The Inheritance of Mental Diseases*. Myerson had been a persistent critic of eugenics since 1913, when he became the clinical director of the Taunton State Hospital in Massachusetts. His first critique against eugenics was published in 1917. That was a critique of studies of the Mendelian basis of mental defects, supported by Davenport and psychiatrist Rosanoff in the United States, as well Ernst Rüdin (more on whom in Chapter 10) in Germany. Davenport and Rosanoff, Myerson wrote, had "worked entirely from the standpoint of Mendelism, and their efforts seem to me to be directed not so much to discover the laws of the transmission of insanity as to fit the facts to Mendelian theory" (something I also pointed out in Chapter 8). He explained that to approach this topic from a Mendelian point of view, it had been necessary to divide mankind into two types: the normal and the neuropathic. Davenport thus considered several varied diseases as due to "the absence of a unit determiner", which made the "neuropathic" constitution considered as a "unit character from a Mendelian point of view".[21]

To these views, Myerson expressed a series of objections. The first one was the assumption that the "neuropathic" differed from the normal by lacking some normal determiner. Myerson noted that there was no evidence for this and that it was equally possible that the disease was due to a diseased determiner or even a new one. It is worth noting that Davenport himself had expressed such a view as early as 1910. While considering cases in which the normally recessive character has become dominant, often described as a reversal of dominance, he noted that they run contrary to the then-current interpretation of dominance, according to which dominance

ABRAHAM MYERSON
1922-1940

FIGURE 9.3   Abraham Myerson (1881–1948). Credit: Unknown. Dr. Abraham Myerson, M'08, Professor and Chairman of Neurology. Tufts University –1940. Tufts Archival Research Center. Medford, Massachusetts. http://hdl.handle.net/10427/003846 (accessed July 14, 2023).

depended on the presence of a determiner and recessiveness on its absence (the presence/absence explanation of Bateson). Davenport suggested that whether a character existed or not was not due to the presence or absence of a determiner; rather, it was due to whether or not it managed to complete its role. He further explained:

> ordinarily, in pure races, a well-developed, pure-bred character has a double determiner as its embryological anlage, while in the heterozygote the determiner is simplex. Now in just these cases when the anlage is simplex the character develops imperfectly; and it is not difficult to understand how, under certain circumstances, the simplex determiner might be insufficient for the development of the organ. Such would result in complete failure to dominate, not a reversal of dominance.[22]

In some cases, therefore, the lack of a dominant phenotype was due to the fact that even though the factor existed in the heterozygote, it did not manage to produce the respective characteristic.

Myerson's second objection was that Mendel's laws had not been shown to apply to single human characters, with the exception of eye color. "To assume then that the vast range of the psychoses (the feeble-minded, the epileptic, character anomaly, criminality, and neuroses) is related to a unit determiner or group of determiners acting as a unit is, to say the least, premature".[23] He also added that normal conditions varied too, and therefore to relate those to a unit was presumptuous, given that the normal condition was "an abstraction". His third objection was about whether "a true Mendelism" had been followed. He noted that the Mendelian ratios emerged from inbreeding among the individuals of the first generation. But this would correspond to the mating of siblings in human families, which was not something that actually prevailed in humans. Therefore, any expectation of Mendelian ratios seemed to be "futile".[24]

Between 1917 and 1925, Myerson wrote more criticisms of eugenics, mostly in the form of book reviews. In those, he criticized the assumptions and the methods of eugenicists. The full-blown attack on eugenics, however, was delivered in his 1925 book *The Inheritance of Mental Diseases*.[25] As previously, he was critical of Davenport's work and conclusions. But most interesting was his critique of Goddard's work, which fell right into his own area of expertise. Myerson first referred to Goddard's 1914 book, with which he aimed to establish "that feeble-mindedness or the liability to become feeble-minded is a Mendelian trait". He noted that Goddard was in agreement with Davenport that 30–40 traits were "neuropathic" due to the lack of a unit character, as well as that Goodard's field investigator and her observations had a key role for their conclusions.[26] That person was Elizabeth Kite, who had come to Vineland in 1909 at the age of 45, having travelled around the world. She had also worked as a teacher of French, which entailed that she was able to read the work of Binet in the original—and actually by 1916 she had translated two volumes of his papers. Having read Binet, however, she was aware that he had insisted on objective standards rather than the subjective judgments of doctors. She was thus aware that Goddard's methods—interviewing family members and deciding whether they were "feebleminded" or not—were violating Binet's methods. Kite justified these violations by relying on judgment, which Binet had considered as the essence of intelligence. She thus asserted that if individuals exhibited good judgment, they should not be classified as "feebleminded". Based on this criterion, she would make judgments in the field about whether mental conditions existed and whether they were the outcome of heredity.[27]

Myerson focused on Goodard's 1912 book on his investigations of the Kallikak family. He noted that the "famous" family "or rather the famous account of the family" had been quoted in all the lay literature and was the key reference for the threat of the predominance of the "feebleminded". "No royal family has enjoyed quite such & prestige as this group", Myerson noted, with the book itself having "all the dramatic flavor of the missionary spirit, or of one who "views with alarm" and wishes to awaken into vigilance the threatened normals".[28]After outlining the story of Martin Kallikak, and of the "good" and the "bad" family lines, Myerson confessed to "a feeling of shame" for the work done by the fieldworker in that case. Having been in charge of a clinic to which "feebleminded" people were brought every day, Myerson was aware that many of those people, who were murderers, thieves, sex offenders, and so on, were in the end shown not to be "feebleminded" at all. In contrast, the mental tests and the psychological examinations had shown that some of those people were "of full average mentality or better". Therefore, he wondered how good the field workers trained by Goddard were to diagnose that a person was "feebleminded" on the spot.[29]

Myerson also questioned the assumptions that the anonymous girl working in the tavern with whom Martin Kallikak had an illegitimate "feebleminded" son, Martin Kallikak Jr., was also "feebleminded". Goddard had written that "after some experience, the field worker becomes expert in inferring the condition of those persons who are not seen, from the similarity of the language used in describing them to that used in describing persons whom she has seen".[30] How is it possible, Myerson asked, to have this kind of knowledge about a girl who lived long ago?

> How can one know anything definite about a nameless girl, living five generations before, of whom there can be no records, whom no one remembers, whom no one has seen? Granting for the sake of argument that perhaps she had a mole on her left arm above the elbow by which she may have been identified by her contemporaries, how can any one know that she was feeble-minded?

If it was possible to have such knowledge, then why the psychological tests that Goddard himself had developed would even be necessary? Myerson compared this to the historical knowledge available for Washington and Lincoln, noting that to acquire this knowledge historians had devoted their lives to study various kinds of sources. But the same could not be said to be the case neither for Martin Kallikak nor for the anonymous tavern girl.[31]

If one was committed to the Mendelian theory, Myerson continued, then one should have the necessary numbers to support their belief, which in turn tempted one to see "feebleminded" people where they might not exist. Furthermore, even if the Kallikak story were true, they would represent an exception rather than the norm. Psychiatrists had seen numerous cases of mental diseases "arising de novo from 'good' stocks", and it was common knowledge that worthwhile ancestors often had failed descendants. Human heredity was still not well understood, Myerson noted, and real life had nothing to do with the Kallikaks. He concluded: "In ethics two wrongs do not make a right, and in science a thousand instances of guess work, intuition, snap judgments and hearsay do good neither to the Mendelian theory nor to eugenics".[32]

In a later section of the book titled "Mendelism and Mental Diseases", Myerson clarified that his criticism was not directed toward Mendel himself or the work of his followers. He did not doubt that unit characters existed and that their patterns of appearance and disappearance could be analyzed mathematically. What he criticized was the "entrance of enthusiastic Mendelians into the field of psychiatry and their conclusions". Even if it were true that these complex mental and physical conditions were unit characters (which himself clearly doubted), the methods by which the data had been collected were questionable. He considered it impossible for anyone who was not medically trained to arrive at conclusions about the mental and physical state neither of dead people on the basis of interviews or written records nor of living people via face-to-face interviews. He would not trust as adequate the field work even of a trained psychiatrist who asked questions to relatives, friends, and enemies. "Such work is not without value, and we must use it at present, but it is too slender to bear the weight of a mighty Mendelian conclusion".[33]

Perhaps the most astonishing fact of all is that by 1927 Goddard himself had changed his mind about the heredity of "feeblemindedness" and had acknowledged that he had been wrong. He noted that in the past he and others had "rather thoughtlessly concluded that all people who measured twelve years or less on the

Binet-Simon scale were feebleminded". But when the war came and measurements were taken of the intelligence of the drafted army, it was found that about 45% of the 1,700,000 soldiers tested did not get above the 12-year limit. Calling all those "feebleminded" Goddard thought in 1927 "was an absurdity of the highest degree".[34] He also admitted that they had been wrong to think of "feeblemindedness" as incurable. Actually, some of those people who had been considered "feebleminded" had been cured and trained to do jobs that were useful.[35] "The problem of the moron is a problem of education. There would be very few, if any, morons in our institutions for the feebleminded if we had not been mistaken in our theories of education", Goddard noted. Now that this was known, the solution was easy. Teach those people what they can learn, rather than those they cannot, and once trained let them be integrated with society and do what they can do.[36]

Despite Goddard's acknowledgment of the faults of his account, in 1942 (!) he published a defense of his Kallikak study, responding to the 1925 criticisms of Myerson.[37] Most importantly, several geneticists continued to accept that mental defects, such as "feeblemindedness" were inherited as Mendelian recessive traits. There were some, however, who were critical of this idea.

GENETICISTS CRITICIZING THE MENDELIAN RECESSIVE ACCOUNT OF "FEEBLEMINDEDNESS"

One of the first critics of eugenics was Thomas Hunt Morgan. Initially, he was a eugenics supporter and also a member of the Committee on Animal Breeding of the American Genetic Association (which until 1914 was the American Breeders Association mentioned in Chapter 8), the society that supported work on practical breeding and eugenics. However, already by 1915, he had become less supportive of the movement. In January 1915 he wrote to Davenport a letter of resignation from the aforementioned committee, complaining about "reckless statements and the unreliability of a good deal that was said".[38] However, Morgan also tried to avoid any unnecessary conflict. He wrote to Davenport that he had no objection as to what people wanted to do but that he did not want to be part of this. However, he gradually took a different stance. In the second edition of his book *Evolution and Genetics*, published in 1925, Morgan included a new chapter that offered "a criticism of the evidence of human inheritance". He justified its addition as being due to "the somewhat acrimonious discussion taking place at the present time concerning racial differences in man, a discussion in which 'nature' and 'nurture' are often confused".[39]

In that last chapter, Morgan considered "feeblemindedness". He noted that until it was better defined when this characteristic began and ended, until it was better understood what internal physical problems produce such a condition or whether it was due to syphilis, it was "extravagant" to claim that there was "a single Mendelian factor for this condition".[40] He also considered insanity and noted that insofar as neither its physiological background nor the external contributing factors were known, its genetic basis was to remain obscure. He added that "The important point, however, to be urged is that the 'mental traits' in man are those that are most often the product of the environment which obscures to a large extent their inheritance, or at least makes very difficult their study".[41] He went on to suggest that he had little faith in the importance of breeding for mental superiority; even though he accepted that there were differences among individuals that were strictly genetic, he thought that for the time being there was no real scientific evidence for this, at least of the kind that was available for animals and plants. The last paragraph of that chapter is worth quoting in full:

> If within each human social group the geneticist finds it impossible to discover, with any reasonable certainty, the genetic basis of behavior, the problems must seem extraordinarily difficult when groups are contrasted with each other where the differences are obviously connected not only with material advantages and disadvantages resulting from location, climate, soil, and mineral wealth, but with traditions, customs, religions, taboos, conventions, and prejudices. A little goodwill might seem more fitting in treating these complicated questions than the attitude adopted by some of the modern race-propagandists.[42]

One geneticist who consistently criticized eugenics was Lancelot Hogben (Figure 9.4). Hogben did not come from an elite social background, in contrast to his close colleagues Julian Huxley and John Burdon Sanderson Haldane. These three became Britain's top biologists and in the 1930s founded, also with Frank Crew, the *Journal of Experimental Biology* and the Society for Experimental Biology. Hogben was largely

FIGURE 9.4   Lancelot Thomas Hogben (1895–1975). Marine Biological Laboratory Black Photo Album 2. Attribution 4.0 International (CC BY 4.0).

self-educated but managed to get a scholarship for Trinity College, Cambridge. During the 1920s he conducted his research on experimental embryology and physiology at the University of Edinburgh, McGill University, and the University of Cape Town. In 1930 he was appointed Chair of Social Biology at the London School of Economics. From this post, he wrote his 1931 book *Genetic Principles in Medicine and Social Science*, which we consider here. In 1937 Hogben moved to the University of Aberdeen and then to the University of Birmingham where he stayed until 1961, with a short interruption during World War II.[43] Historian and philosopher of science James Tabery has studied Hogben's archive in detail and has found two remarkable features in Hogben's reminiscences: his animosity for eugenicists such as Fisher and his lack of serious mention of his publications attacking the British eugenics movements, which Hogben simply considered as the moral obligation of his chair of social biology at LSE.[44] According to historian Pauline M.H. Mazumdar, Hogben was an active and committed socialist, a fact of which he himself had lost sight of 60 years later, when he wrote his autobiography. Instead of joining the Eugenics Society, as Fisher did, in 1914 he joined the Cambridge University Fabian Society, becoming its secretary, which resulted in the Fabian Society's politics moving sharply to the left.[45] It should be noted though that many Fabians were enthusiastic eugenicists.[46]

In his book *Genetic Principles in Medicine and Social Science*, right before quoting Marx, Hogben wrote that people who paid exclusive attention to the genetic aspect of a condition tended to attribute a lot to inborn racial differences. This was a prejudice that would not characterize the study of animal behavior. However, humans had a culture that differentiated them from animals. *"Thus increasing complexity of cultural achievement may proceed in human societies independently of any change in man's inborn equipment"*.[47] Hogben continued by noting that his viewpoint made no assumptions about inborn changes in humans. There was no reason to think that humans of his time were different from humans living in the New Stone Age, any more than Europeans living in 1750 were different from Europeans living in 1850. His viewpoint did not deny that there were differences that distinguished humans at the individual level. As he put it:

> a mutation or unusual gene combination is not likely to occur once and once only, or to occur in one country and not in another. In a statistical sense it may well be that inborn differences of temperament do distinguish different racial groups. We have at our disposal no precise information about such differences at present.[48]

Hogben had tried to accommodate the standpoint of dialectical materialism, "the official philosophy of the Communist movement", as he put it, in his own system of thought. The influence of this was his interaction with Soviet scientists, such as Nikolai Bukharin during the Second International Congress on the History of Science, which took place in London in 1931. The Marxist influence on Hogben's thinking about the problems of social biology sharpened his perception of the class-bound nature of the eugenic program and provided a theoretical background for his campaign against the overemphasis of biology in human society.[49] However, it seems that perhaps Hogben did not really need the Marxist framework for arguing against the overemphasis of the influence of biological factors on society, as he had relied—as I also show below—on a solid scientific basis to develop his arguments.[50]

Here is what Hogben had to say about "feeblemindedness" in his 1931 book. He began by considering the various types of mental defects in the English law, which distinguished among "idiocy", "imbecility", and "feeble-mindedness". After considering their legal definitions, Hogben noted: "Doubtless these definitions reflect the utmost credit on the intelligence of the legal profession. They have no relevance to biological inquiry".[51] He suggested that the clinical classification of mental cases was much more comprehensive, cutting across the legal distinction between idiots, imbeciles, and feebleminded. The clinical categories were so many and varied, and any of those might include individuals distinguished as idiots, imbeciles, or feebleminded. Acknowledging that the classification of mental ability was based on psychological tests, Hogben noted that, on the one hand, performance in psychological tests could be investigated from the perspective of genetics. On the other hand, the distinctions made between the "normal" and "defective", as well as among "defectives", were arbitrary. He thus concluded that "it is therefore unjustifiable to hope that the genetic aspect of feeble-mindedness can be investigated by the simple procedure appropriate to the study of clear-cut differences like coat colour, plumage, or the shape of the comb in poultry".[52] Simply put, behavioral characteristics in humans were not as clear cut as some traits typically studied in Mendelian genetics. Therefore, there was no justification for applying Mendelian genetics rules to the study of "feeblemindedness".

Hogben then argued that the consideration of the various ways that mental defects are classified made necessary the consideration of two conclusions that were often overlooked in discussions about their heredity: "The first is that *feeble-mindedness is clinically not one thing but many things. There is no one single problem of the hereditary basis of feeble-mindedness*". The second was that "So long as any group of defectives is differentiated from the normal by reference to the results of intelligence tests, *it is impossible to entertain the likelihood that a single gene is involved in the distinction*".[53] Hogben went on to explain why this was the case. The scores of intelligence tests in the population formed a normal distribution. However, such a distribution could be the result of two conditions. On the one hand, it could be due to the interactions of the same genotype with variable external environments. On the other hand, it could also be the cumulative outcome of many such curves, each describing the range of developmental variation of a different genotype. Therefore, if the variability of intelligence was largely determined by genetic differences, there was no way that an arbitrarily selected segment of the normal curve of variation could separate a single genotype from the rest of the population. Hogben concluded: "To put the matter in another way, *the more we emphasise the genetic basis of variability in the scores of intelligence tests, the more are we committed to regard any form of feeble-mindedness defined by psychological tests as a genetically complex group*".[54]

Hogben then turned to a 1907 study of the relatives of asylum inmates by David Heron, Second Galton Research Fellow in National Eugenics, working at the Francis Galton Laboratory for National Eugenics at the University of London. Heron had compiled the data presented in Table 9.2. These results, Heron wrote, compelled one to consider sanity as dominant. He thus used the symbol *D* for the respective factor and the symbol *R* for the factor related to insanity. The observed ratios for

TABLE 9.2

Incidence of Insanity in Parents and Offspring in a Study of the Relatives of Inmates of Asylums, and the Expected Incidence Based on Mendelian Genetics

Nature of Parents as to Insanity	Total Offspring Recorded		Percentage Offspring Recorded		Percentage Offspring Expected Based on Mendelian Genetics	
	Insane	Sane	Insane	Sane	Insane	Sane
Both insane (RR × RR)	4	4	50	50	100	0
One insane (DR × RR)	93	299	24	76	50	50
Both sane (DR × DR)	314	1179	21	79	25	75

Source: Based on Tables IV and V in Heron (1907), p. 17.

those cases in which both parents were sane, 79% sane and 21% insane, seemed to approach well the expected ones based on Mendelian genetics, 75% sane and 25% insane, or 3 sane:1 insane. However, there were significant deviations in the other cases. The observed ratios for the cases with one parent sane and one parent insane were 76% sane and 24% insane, whereas they should have been expected to be 50% sane and 50% insane, or 1 sane:1 insane. For those cases with both parents insane, the observed ratios were 50% sane and 50% insane, whereas they should have been expected to be 100% insane. As Heron wrote,

> No argument in favour of Mendelism seems possible on these figures . . . there is no obvious reason why the expectation should be so much more closely fulfilled by the record in the third mating than in either of the other two.[55]

Also quoting Heron and commenting on these results, Hogben wrote:

> It would be more correct to say that we cannot explain these figures by any Mendelian hypothesis which is applicable to characteristics which manifest themselves in almost any environment in which the organism will live. On the other hand, there is no prima face reason to suppose that any form of insanity is of such a nature, especially in view of the fact that insanity may develop comparatively late in life.[56]

We see here not only that a fellow of the Francis Galton Laboratory for National Eugenics in London, working under Pearson, came across results that did not conform to Mendelian genetics but also that Hogben developed a purely scientific critique, notwithstanding his political views. It should be noted that Heron was an ardent eugenicist, concerned that how eugenics was done on the U.S. side and their errors would not support the common effort.[57]

Subsequently, Hogben considered data about the high incidence of insanity among first-borns. After considering various datasets and their characteristics, he compared those data to data about the relatively high incidence of "idiots", still-births, and deaths during the first year among first-borns. His conclusion was that with respect to insanity and dementia, the study of environmental factors ("including prenatal nutrition, difficulty of labour, lactation, toxemias of pregnancy, maternal diet, and the social environment as determined by the size of family") required investigation for them to be controlled or eliminated. And he added:

> We can say in addition that whatever genetic differences enter into the determination of such conditions are not likely to be understood so long as we neglect the demonstrable significance of environmental agencies in the determination of "mental" disorders and defects.[58]

The more was known about the role of the environment, Hogben suggested, the better we could understand the genetic aspect of the problem.

Hogben concluded Chapter II of his book by stating:

> There is now justification for believing that genetic factors which play a part in the aetiology of feeble-mindedness and insanity will prove to be incompatible with the theory of Particulate Inheritance when we know something about them. There is still less justification for such conclusions as those which Davenport and Goddard have put forward in their studies on insanity and feeble-mindedness. In so far as the objections directed by Heron against such work are based on fact, they are to be regarded as a contribution to the advancement of scientific knowledge, whatever views one may entertain concerning the theoretical bias of the author.[59]

What made Morgan and Hogben perceive the situation differently than East, Punnett, Fisher, and the other eugenicist-geneticists? One can attribute this to their personal political, social, or other views. However, there is another difference that might be the most important one: both Morgan and Hogben were experimental embryologists, who had a deep understanding of the complexities of development. Morgan had started his career as an embryologist and it seems that he maintained his interest in it ever since, returning to it with a 1934 book on the topic. Hogben also conducted important experimental research in embryology, especially after 1919, and he was fundamental in establishing experimental zoology as a discipline in England. Also, for both of them evidence came first, whereas they disliked speculation. So rather than adopting a model, as the Mendelians did, and trying to find evidence to support it, or stretching it to accommodate exceptions, they followed the evidence and drew conclusions from it. This is why Morgan underwent crucial shifts in thought around 1910, as we saw. He followed the evidence and was ready to reject his previous views. So did Hogben too, who neither hesitated to argue against their friends when speculation went too far.[60] Here is then an important lesson from history: those who rejected naive Mendelian genetics and eugenics had a deep understanding of development, thanks to their experimental work, and prioritized evidence over speculation. Isn't this something that could inform school teaching? I return to this point in the Conclusions.

FIGURE 9.5　John Burdon Sanderson Haldane (1892–1964). John Innes Archives courtesy of the John Innes Foundation.

A belated critic of eugenics, close to Hogben politically, professionally, and personally, was John Burdon Sanderson (Jack) Haldane (Figure 9.5). He came from a family with perhaps extreme scientific interests and practices, his father being John Scott Haldane, a physician and physiologist who had made several important discoveries about the human body. The extremeness had to do with the fact that these discoveries were due to experimentation that Haldane Senior did on himself or on his son. Jack was educated in Oxford; he received a degree in classics, went away to fight in World War I, and got back to focus on evolution and genetics. He was never really a good experimentalist with respect to practical matters in the lab. But he had a gift for numbers, and this drove all his scientific accomplishments. As a result, he did not do any lab work himself; rather, he drew on the experimental data of others where he identified problems or figured out patterns. His first major paper in 1922 was the one in which he presented what is described as Haldane's Rule: the observation that when in the offspring one sex is absent, this is the heterogametic sex, that is, the one that has two different sex chromosomes such as X and Y. Then, starting in 1924, he published nine papers in ten years that contributed to the development of what became known as population genetics. It was around that time that he was open to the ideas of eugenicists. Whereas he insisted that nobody had proved that black people were less intellectual than whites, he did not rule out any connection between race and intelligence and he did doubt that Africans could ever reach the level of Europeans. When he first spoke about eugenics in 1923, what he had in mind was not the coercive measures of the day, but some sophisticated manipulations that he thought humans someday would be able to apply.[61]

On April 19, 1928, he gave the Conway Memorial Lecture at Essex titled "Science and Ethics". At the time he accepted some of the assumptions of eugenicists. "Feeble-mindedness is fairly strongly inherited, but unfortunately it is generally inherited in such a way that the segregation or massacre of the feebleminded, even if continued for several generations, would not stamp it out", he argued. He noted however that the "feebleminded" did not necessarily produce "feebleminded" offspring, unless they mated with one another. "If, therefore, the feeble-minded are to be segregated, it should be in their own interests, and because they are unfit to bring up a family, quite as much as on eugenical grounds".[62] He also admitted that it seemed likely that there was "a correlation between wealth and the hereditary factors determining intelligence, because the well-to-do include many families of the professional classes in whom intelligence is undoubtedly hereditary, and the unskilled labourers include the majority of the feebleminded". He added that a lot was still unknown about the inheritance of intelligence in order to predict the outcome of selection after a few generations. But if we grant the case of the extreme eugenist, what is the remedy?" he asked. If eugenicists asked to penalize the very poor for having large families, or cut down the financial support for them, then there was "a serious conflict between science and the dictates of the conscience of most enlightened men and women.[63] He also added that a lot that he considered as both unscientific and immoral had been advocated in the name of eugenics.[64]

However, in a 1938 book, titled *Heredity and Politics*, Haldane made a full-blown attack on eugenics. He began by criticizing the science, noting that the vast majority of recessive genes, which were the ones that eugenicists wanted to eliminate with sterilization, were found in heterozygotes. Natural selection would not have any effect on them, but only on the homozygotes. But their frequency depended on the amount of inbreeding in the population, and that was a social issue to address. He also added that the frequency of "albinism, recessive types of blindness, and various types of idiocy" would surely and slowly increase in human populations. However, he noted that was not an urgent problem, as the increase would take thousands of years. And he added: "We need not worry about the approaching degeneracy of the human species. We have at the present moment other and more urgent problems to consider".[65] He also explained the social dimensions of the problem:

> At the present time nearly all cases of lunacy or gross idiocy are probably reported in all social classes. However, there is reason to think that mild mental defect is much less frequently certified among the rich than the poor. A well-to-do family can afford to keep a "backward boy" or "a girl who was no good at school". A poor family cannot. Sterilization of all certified defectives would thus in our society be a class measure.[66]

Haldane added that "mental defect is to a large extent a social rather than a biological problem". He explained that in a society where there would be work for everyone, it would be possible for those who were considered "feebleminded" to find something to do. But this was not the case in present society, and he suggested that it might be due to their difficulty in finding a regular job under the given circumstances that those people were certified as feebleminded, rather than due to the fact that they were of low intelligence. Therefore, this might in fact be "a social and economic rather than a biological phenomenon". As a person Haldane knew preferred "feebleminded" men to look after his pigs, he argued: "I am of the opinion that a man who can look after pigs or do any other steady work has a value to society, and that we have no right whatever to prevent him from reproducing his like".[67]

A note of caution is necessary here. Whereas in this chapter I have presented geneticists as belonging into two broad groups, those accepting that "feebleminded-ness" was a Mendelian recessive trait and those who did not, one should refrain from thinking of these two groups as comprising proponents and opponents of eugenics, respectively. It is indeed the case that some argued clearly against the simplistic Mendelian genetic accounts underlying eugenics, such as "feeblemindedness" being Mendelian recessive, whereas others did not. However, as historian Diane Paul has noted if one considers their writings more carefully virtually all geneticists, includ-ing those who were considered and considered themselves fierce critics, endorsed *some* kind of eugenics.[68] We saw that Morgan and Haldane up to some point did not disagree with all eugenics ideas. Even Hogben, who was certainly critical of many specific aspects of the eugenics movement of his day, was not by any means opposed to eugenics in principle. For instance, in his book *Genetic Principles in Medicine and Social Science* that we considered earlier, he wrote:

> Eugenics was defined by Galton as the study of agencies under social control which may improve or impair the racial qualities of future generations. With such a proposal it is difficult to see that any reasonable person would disagree. Writers on economics sometimes distinguish between "positive" and "negative" eugenics. Negative eugenics is simply the adoption of a national minimum of parenthood, an extension of the prin-ciple of national minima familiarised in the writings of Sidney and Beatrice Webb. It is thus essentially en rapport with the social theory of the collectivist movement.[69]

ON SCIENTISTS BEING EUGENICISTS

It is in fact the case that several scientists were ambivalent and inconsistent with respect to eugenics views. An interesting example is Raymond Pearl (Figure 9.6), who not only evolved from a strong proponent of eugenics in the 1900s to a fierce critic in the 1920s, but he was also incoherent with respect to his antieugenic crit-icisms. Pearl was Professor of Biology in the Medical School and in the School of Hygiene and Public Health of the Johns Hopkins University. After finishing his PhD at the University of Michigan in 1902, he became interested in the applications of statistical methods to biological problems and during the year 1905–1906 he worked with Karl Pearson at University College, London, also joining the editorial board of *Biometrika* in 1906. However, Pearson removed Pearl from the editorial board of the journal in 1910 because he began accepting Mendelism and rejecting the Law of Ancestral Heredity.[70] Pearl returned to the United States in 1907, working for a year at the University of Pennsylvania and then until 1918 at the University of Maine at Orono. His contributions during those years were mainly to the applications of

FIGURE 9.6 Raymond Pearl (1879–1940) between 1910 and 1920. Library of Congress, Prints and Photographs Division [LC-DIG-npcc-30766].

statistical methods in biological research. He was thus invited in 1918 to become Professor of Biometry and Vital Statistics in the newly founded School of Hygiene and Public Health of Johns Hopkins University. He was a prolific author of 712 titles, including 17 books, and also the founding editor of the journals *Quarterly Review of Biology* and *Human Biology*.[71]

Perhaps not surprisingly given his connection to Pearson, Pearl was also initially a proponent of eugenics. Soon after returning to the United States, he wrote a paper titled "Breeding Better Men: The New Science of Eugenics Which Would Elevate the Race by Producing Higher Types".[72] The title is indicative of Pearl's support for eugenics, I think. In 1912 he also contributed a paper to the volume that emerged from the First International Eugenics Congress, which took place in London, in which he presented findings on the fecundity in domestic fowl. His main conclusion was that fecundity in the domestic fowl was inherited "strictly in accordance with Mendelian principles". He also noted that the observed variation in this trait was due to two separately inherited factors, L_1 and L_2 (the latter being sex-limited), and that high fecundity required the presence of both of these factors in the same individual. Otherwise, if only one of these factors were present, a similar degree of low fecundity was manifested. Finally, he noted that there was a "definite and clear-cut segregation of high fecundity from low fecundity".[73] We see here once again the familiar pattern of Mendelian genetics accounting for phenotypic traits first established by Bateson. By invoking the necessary number of factors, one can account for the Mendelian inheritance of phenotypic traits.

But what fecundity in fowls had to do with humans and eugenics? Pearl explained:

> While these results may have no direct eugenic bearing, they do, I believe, have an important indirect connection with eugenic problems. In the first place, these results furnish a novel conception of the mode of inheritance of fecundity. They show that this highly variable physiological character is inherited in accord with simple Mendelian principles. They further show that simple selection of highly fecund females alone is not sufficient to ensure high fecundity in the race. From the eugenic standpoint they suggest, though of course they do not prove, that possibly some part of the observed decline in human fecundity in highly civilized races may be due to the dropping out or loss of one or more of the genes upon which high fecundity depends, this loss being coincident with the complete cessation of the natural selection of highly fecund types.[74]

This is Mendelian "eugenics" at its best. Pearl went on to suggest that if by studying the inheritance of phenotypic traits in vertebrates it is concluded that they were inherited in a simple Mendelian fashion, this should be considered as indicating that something similar might be the case for humans, thus encouraging a similar approach to the study of human phenotypic traits. Similar arguments were made in another paper published in the same year, where Pearl argued that the progress of "scientific eugenics" must rely on the experimental study of hereditary phenomena in plants and animals. "Genetics is at once the guide and the support of eugenics", he concluded.[75]

It was however the very same person who in 1927 published one of the first critiques of eugenics. It was titled "The Biology of Superiority", and it was published in *The American Mercury*, a widely read magazine published by Pearl's close friend, Henry Louis Mencken. After reviewing the origin of eugenics, and the biometric and Mendelian approaches to the study of heredity, Pearl concluded that human heredity was no different than that of animals and plants. He accepted that some human traits were inherited in a Mendelian recessive fashion—presenting blue eye color as an example (even though as we saw there was evidence since 1908 that this might not be the case). Then he wrote:

> Eugenics has fallen in some degree into disrepute in recent years because of the ill-advised zeal with which some of its more ardent devotees have assigned such complex and heterogeneous phenomena as poverty, insanity, crime, prostitution, cancer, etc., to the operation of either single genes, or to other simple and utterly hypothetical Mendelian mechanisms. But discounting all such stupidity, because in the long run it is certain to have only its just effect upon the progress of human biology, the solid achievements of critically scientific eugenics up to the present time are unquestionably considerable.[76]

Without outrightly rejecting eugenics, Galton or Pearson, we see Pearl criticizing the exaggerated application of "Mendelian" eugenics: accounting for complex human traits and behaviors in terms of simple Mendelian genetics, something that he described as "stupidity". Whereas propaganda and science were closely linked since the beginning of eugenics, Pearl continued, in recent years they had become so closely intertwined transforming the eugenics literature to "a mingled mess of ill-grounded and uncritical sociology, economics, anthropology, and politics, full of emotional appeals to class and race prejudices, solemnly put forth as science, and unfortunately accepted as such by the general public".[77] The problem for Pearl was that propaganda had prevailed. Its two key characteristics are serving the interests of those who produce it and its indifference to the truth; the outcome had been the production of eugenics propaganda carried out in the name of science.

After presenting the main theses of eugenics, Pearl noted that they were based upon, and derived their meaning from, a profound fallacy: "that the essence of heredity is comprehended in the statement that like produces like".[78] However, he noted this was not what research in genetics had shown, quoting from Morgan's 1925 book as I also did earlier. Pearl suggested that the current knowledge about heredity made clear that any attempt to predict the traits of the offspring from the respective traits of the parents was doomed to fail. Therefore, a reconsideration of the theses of eugenics was necessary. Without rejecting the slogan of positive eugenics "to breed better men" he himself had used years ago, Pearl suggested it was necessary to reconsider how this could be done, and one way to do this would be to consider how many superior men also had superior offspring. And this is what he did. Drawing on Encyclopedia Britannica that had over 25,000 biographical notes, he looked for cases of parents and offspring to whose biographies one whole page or more of space was given. He thus arrived at a list of 63 philosophers and 85 poets of great eminence. What Pearl found was that in most cases their parents and their children were not equally eminent. Based on these results, he concluded:

> A new ad hoc investigation of the breeding of great men shows that the facts are in full accord with the expectation from established genetic principles, and not at all in agreement with current eugenic dogma. It would seem to be high time that eugenics cleaned house, and threw away the old-fashioned rubbish which has accumulated in the attic.[79]

What turned Pearl from a proponent to a critic of eugenics is unclear. However, what he seems to have rejected in 1927 was not the goals of eugenics, but the basic theses about how these goals would be achieved. He criticized the oversimplifications and exaggerations of "Mendelian" eugenics. He did not reject the racism underlying eugenics, but its scientific disguise. Most interestingly, he held himself anti-Semitic views that co-existed with his public criticisms of eugenics. In a letter of June 1922 to his friend Harvard Professor Lawrence Joseph Henderson, after the latter's university had tried to impose a quota on Jewish students, Pearl wrote: "I think we do it much more neatly here. We make no noise about it, but, for a number of years past, very quietly and skillfully means have been taken and are being planned for the future to keep down our Jewish percentage".[80] As Henderson was quoted saying in Pearl's obituary: "Throughout his life he felt himself a north of Boston man and cultivated and cherished the sentiments and some of the prejudices of his people".[81]

This is why simple divisions among opponents and proponents are not easy to make. Especially, when it comes to opponents of eugenics, one should carefully consider what exactly they were against. Criticisms can be scientific, ideological, or both. Science can be invoked to support ideology, and ideology can be invoked to drive science. But these can be combined in various ways. As historian Elazar Barkan put it:

> scientific and cultural values influence each other. Yet one should not *a priori* assume that the correlation between the political and the cultural position of scientific ideas is rational. Incoherence and contradiction are more likely—and Pearl is a case in point.[82]

The ideology of eugenics would become the basis for much harm against the Jewish population of Europe. In the Preface of his 1938 book *Heredity and Politics*, Haldane expressed his concern about what was happening in Germany at the time:

> The stringent measures which have been taken in Germany, both for the expulsion of Jews from many walks of life, and for the compulsory sterilization of many Germans, are said to be based on biological facts. I do not believe that our present knowledge of human heredity justifies such steps.[83]

It is now time to see what was happening in Germany that made Haldane write this, and what was its cause.

NOTES

1 I should note that I am aware that not all the people mentioned in Table 9.1 were geneticists in the strict sense; nor would they have identified themselves as such. However, because they were all doing research that had implications for genetics, I use the term "geneticist" as a shorthand description here.
2 Barker (1989), p. 362.
3 Jones (1944).
4 East (1912) pp. 137–138.
5 East (1917), pp. 216–217.
6 Punnett (1911), p. 10.
7 Punnett (1912), p. 137.
8 Punnett (1917), p. 465.
9 Punnett (1952). pp. 346–347.
10 Bateson (1908), pp. 30–32.
11 Bateson (1909), p. 304.
12 Bateson (1921/1922), pp. 326–327.
13 Bodmer et al. (2021).
14 https://profjoecain.net/what-is-wrong-ronald-aylmer-fisher/ (accessed June 8, 2023).
15 https://thetab.com/uk/cambridge/2020/06/25/caius-college-decides-to-remove-commemo-rative-window-to-eugenicist-alum-139190 (accessed June 8, 2023).
16 Fisher (1924), p. 114.
17 Fisher (1924), p. 115.
18 Fisher (1924), p. 115.
19 Paul (1998), p. 122.
20 Lippmann (1922).
21 Myerson (1917), pp. 359–360.
22 Davenport (1910), p. 131.
23 Myerson (1917), p. 360.
24 Myerson (1917), p. 361.
25 Trent (2001).

26 Myerson (1925), p. 77.
27 Zenderland (2001), pp. 159–161.
28 Myerson (1925), p. 77.
29 Myerson (1925), p. 78.
30 Goddard (1912), p. 15.
31 Myerson (1925), p. 79.
32 Myerson (1925), p. 80.
33 Myerson (1925), p. 279.
34 Goddard (1927), p. 42.
35 Goddard (1927), p. 44.
36 Goddard (1927), p. 45.
37 Goddard (1942).
38 Quoted in Allen (1978), p. 229.
39 Morgan (1925), p. vi.
40 Morgan (1925), p. 201.
41 Morgan (1925), p. 203.
42 Morgan (1925), p. 207.
43 Wells (1978). George Philip Wells had been a student of Hogben's department in London; he was the son of H.G. Wells (see Chapter 8).
44 Tabery (2006).
45 Mazumdar, (1992), p. 109.
46 Paul (1998), Chapter 2.
47 Hogben (1931), pp. 168–169 (emphasis in the original).
48 Hogben (1931), p. 169.
49 Mazumdar, (1992), pp. 114–116.
50 Erlingsson (2016).
51 Hogben (1931), p. 35.
52 Hogben (1931), p. 97.
53 Hogben (1931), p. 110 (emphases in the original).
54 Hogben (1931), p. 110 (emphasis in the original).
55 Heron (1907), p. 17.
56 Hogben (1931), p. 111.
57 Paul (1998), p. 119.
58 Hogben (1931), p. 116 (emphasis in the original).
59 Hogben (1931), pp. 171–172.
60 Maienschein (2016); Erlingsson (2016).
61 Subramanian (2020), Chapter 3.
62 Haldane (1928), p. 23.
63 Haldane (1928), p. 25.
64 Haldane (1928), p. 29.
65 Haldane (1938), pp. 73–74.
66 Haldane (1938), p. 98.
67 Haldane (1938), pp. 100–101.
68 Paul (1995), Chapter 1.
69 Hogben (1931) pp. 209–210.
70 Barkan (1992), p. 154.
71 Jennings (1942).
72 Pearl (1908).
73 Pearl (1912a), pp. 56–57.
74 Pearl (1912a), p. 57.
75 Pearl (1912b), p. 339.
76 Pearl (1927), p. 260.

77 Pearl (1927), p. 260.
78 Pearl (1927), p. 261.
79 Pearl (1927), p. 266.
80 Quoted in Barkan (1992), p. 215.
81 Jennings (1942), p. 308.
82 Barkan (1992), p. 219.
83 Haldane (1938), p. 9.

10 Mendelian Genetics and the Nazi Racial Hygiene

U.S. RACISM MEETS NAZI RACIAL HYGIENE

What might connect Mendelian genetics to the Nazi racial hygiene? The link is the eugenics movement in the United States that was based on Mendelian genetics and that was often an expression of racist views. Contrary to what one might expect, between the eugenicists in the United States and their German counterparts there was a long-term, close relationship of mutual admiration. One might say that the connection began with the First International Eugenics Congress, which took place in London in 1912. This was the meeting where Punnett presented the paper "Genetics and Eugenics" we considered in the previous chapter, in which he argued that "feeblemindedness" was a case of simple Mendelian inheritance that could be eliminated by strict segregation.[1] Bateson was listed among the delegates, representing the Entomological Society of London and the Linnean Society, as was Punnett. The meeting was organized by the Eugenics Education Society, and its aim was to give the opportunity to those engaged in the scientific study of eugenics to meet and exchange views. Why? Because as Leonard Darwin, son of Charles Darwin and president of the conference, wrote in his introduction to the meeting book:

> Eugenics, as Sir Francis Galton termed the study of the agencies under social control that may improve or impair the racial qualities of future generations, presents, it was stated, problems of the utmost social importance. At present the most urgent need is for more knowledge, both of the facts of heredity and of the effects of social institutions in causing racial change. As knowledge accrues, it must be disseminated and translated into action. The imparting of such knowledge would constitute a great advance in education: for both private individuals and public bodies have yet to be impressed with the gravity of the situation, and induced to act on eugenic principles. Ultimately it may be possible to induce Society to adopt a well-considered eugenic policy and to carry out reforms on eugenic lines.[2]

This was the gospel. Apply science for the common good, and to do this, it was necessary to bring together the "experts".

Not surprisingly, Davenport was among the experts presenting at this conference. His paper titled "Marriage Laws and Customs" explained that there was a biological justification for the prohibition of marriages between close relatives, as well as between people who had mental conditions. However, he added, any laws against the marriage of the "feebleminded" were "futile", as they would anyway find another

DOI: 10.1201/9781032449067-12

"feebleminded" person and have children together. "It would be as sensible to hope to control by legislation the mating of rabbits. The only way to prevent the reproduction of the feeble-minded is to sterilize or segregate them", Davenport concluded. The last legal limitation he considered was about marriages that resulted in the "mixture of races".

> The biological basis for such laws is doubtless an appreciation of the fact that negroes and the other races carry traits that do not go well with our social organization. For the Ethiopian has not undergone that selection that in Europe weeded out the traits that failed to recognize property rights, or that failed to give industry, ambition and sex control.[3]

Many disasters would be avoided if the law took lessons from biology, Davenport concluded.

Davenport was listed among the vice-presidents of the conference, along with some very prominent figures such as Alexander Graham Bell, Founder of the Volta Bureau; The Right Hon. Winston Churchill, M.P., First Lord of the Admiralty; David Starr Jordan, Principal, Leland Stanford University, President of the Eugenic Section, American Breeders' Association; and August Weismann, Professor of Zoology, Freiburg (which I have not been able to explain; Weismann had just retired and died two years later). Among these vice-presidents there was also Alfred Ploetz, President of the International Society for Race Hygiene in Germany, and also President of the German Consultative Committee that included among its members Eugen Fischer and Ernst Rüdin.

Ploetz was a physician and the person who had coined the term racial hygiene ("Rassenhygiene"), the German equivalent to "eugenics", in his 1895 book *Die Tüchtigkeit unsrer Rasse und der Schutz der Schwachen: ein Versuch über Rassenhygiene und ihr Verhältniss zu den humanen Idealen, besonders zum Socialismus* (*The Fitness of Our Race and the Protection of the Weak: An Attempt on Racial Hygiene and Its Relation to the Human Ideals, Especially to Socialism*). Given what I have presented in the previous chapters, it may come as a surprise that it was Ploetz who founded the first eugenics-focused journal in the world, the *Archiv für Rassen—und Gesellerschaftsbiology* (*Journal of Racial and Social Biology*) in 1904. It was also Ploetz, along with others including his future brother-in-law Rüdin, who established in 1905 the first professional eugenics organization, the Society for Racial Hygiene. So both the first eugenics journal and the first eugenics society in the world were founded in Germany. For comparison, the British Eugenics Education Society was founded in 1907, and the Eugenics Records Office in the United States was founded in 1910.[4] This is important to keep in mind because the influence of U.S. eugenics on Nazi racial hygiene should not be taken to imply that a eugenic line of thought did not already exist in Germany.

Racial hygiene took a Mendelian turn early on. The first volume of Ploetz's *Archiv* included an essay by Correns on experimental Mendelian research. Tschermak also published in the same journal in 1905 an essay on Mendelism and Galton's theory of ancestral heredity. Fischer, an anthropologist, and Rüdin, a psychiatrist, had corresponded with Davenport for a few years before the 1912 London conference. Importantly, they were both pioneers in the introduction of Mendelism in their fields.

Fischer famously studied the so-called Rehobother Bastards (also called *Basters*), which were the offspring of Dutch settlers and the native Hottentot women in Southwest Africa. He began the study in 1908 and by 1910 he had attempted to apply the typical Mendelian model to this case. Whereas his results were not clear cut, there were cases that seemed to fit that model. His study was eventually considered the first successful demonstration of Mendelian genetics in human populations. Rüdin also noted in 1908 that Mendelian reasoning about mental disorders was generally absent from the relevant studies until that time. He thus attempted to investigate this, and by 1911 he had arrived at the conclusion that certain forms of schizophrenia were inherited as Mendelian recessive traits, where many psychopathic conditions were inherited as dominant.[5] Fischer and Rüdin eventually became two of the most prominent scholars in the Nazi academic world. Fischer became the director of the Kaiser Wilhelm Institute of Anthropology, Human Heredity, and Eugenics in Berlin in 1927. Rüdin became director of the Department of Genealogical and Demographic Studies at the German Institute for Psychiatric Research, established in 1917 in Munich, which in 1924 also came under the auspices of the Kaiser Wilhelm Society. Rüdin became director of the whole institute in 1931.[6]

There were other people due to whom German racial hygienists were informed about eugenics in North America, such as Géza von Hoffmann, who had been the Austrian vice-consulate in California for several years. Hoffman seems to have been the main contact between German and U.S. eugenicists until the late 1910s, although the outbreak of World War I made communication difficult. In 1913, he published a book titled *Racial Hygiene in the United States of North America*, where he reported the widespread acceptance of eugenic ideals in the United States. A large section of the book was devoted to sterilization that in his view was the easiest means to prevent the reproduction of inferior people. In 1914, Hoffmann reported to the journal of the International Society for Racial Hygiene a proposal of the American Genetic Association that he found very radical and perhaps unfeasible, but one he praised for being in the right direction.[7] The report was written by Harry Laughlin, Davenport's assistant director at the Eugenics Record Office, and the proposal is so unbelievable that is worth quoting in full:

> The passage of the model law recommended in this chapter, if supported by ample appropriation, and vigorously and competently enforced, will, the committee is confident, inaugurate in any state so undertaking it an eugenical program which, if consistently followed and supported by all of the remaining states, will in two generations practically cut off the inheritance lines and consequently the further supply of that portion of the human stock now measured by the lowest and most degenerate one-tenth of the total population. This portion of the American breeding stock now constitutes a growing menace to the conservation of our best blood, and consequently to the very foundations of our social welfare. Only those nations of history which have arisen to great effort to achieve a worthy end have long enjoyed a high plane of culture or have left contributions of worth to subsequent peoples. The purging of defective traits from the blood of the American people is worthy of our best efforts.[8]

Several people in Germany wanted it to be a nation that would make that great effort. But the inspiration, the insights, and the previous experience and knowledge came from the United States.

Because of World War I, the international relations among eugenicists became strained. The second International Congress for Eugenics took place at the American Museum of Natural History in New York in 1921, without the participation of Germans. For many in Germany the fact that it was blamed for the war and that former German territories were occupied by the French and the Belgians, precluded them from participating in any international events. Davenport was concerned about this, and so tried to bring the Germans back to the international eugenics movement. As Davenport wrote to geneticist Fritz Lenz on July 13, 1923, "There is no country which has higher ideals [with respect to eugenics] than Germany and we assume that she will assume a leading position at the next congress".[9] Lenz was another key figure in Germany, having completed a dissertation on sex determination and inherited disease in 1912, who also put racial hygiene on a Mendelian basis. Eventually, by 1925 the Germans joined the international eugenics movement anew and the relations with others had largely been restored. By that time, it was Lenz who had become the main contact between the United States and Germany, having established excellent relations with both Davenport and Laughlin. This close relationship was also characterized by financial support. Several U.S. foundations supported the work of German eugenicists, most importantly the Rockefeller Foundation in New York. This support continued even after the Nazis came to power in 1933.[10]

There was also continuous influence coming toward Germany from the United States. In 1929 a book titled *Sterilization for Human Betterment: A Summary of Results of 6,000 Operations in California, 1909–1929* was published that both influenced and helped legitimize the German sterilization law. The authors were two eugenicists: Ezra Gosney, a philanthropist and founder of the Human Betterment Foundation, and Paul Popenoe, a family relations researcher and agricultural explorer. In their book they reported that by January 1, 1929, California had performed 6,255 sterilizations, which they described as the longest continuous record of sterilization compared to any other state in the world. They also mentioned that by July 1, 1929, 22 states actually had sterilization laws; however, in most cases either these laws had not been put into effect or the number of operations performed was very small.[11] In their introduction, they noted that "Eugenic sterilization of the hereditary defective is a protection, not a penalty, and should never be made a part of any penal statute", adding that the U.S. Supreme Court had confirmed the legality of eugenic sterilization in the case of *Buck vs. Bell* in which Justice Holmes, as we saw in Chapter 8, made the famous statement: "Three generations of imbeciles are enough".[12] A key argument in their book was that the principal reasons for sterilization were eugenic. People should be sterilized, they argued, if it was for the common good not to bear any offspring, and sterilization had been shown to be the most effective and satisfactory means to achieve this. They thus suggested that sterilization was justified "(1) if mental disease and defect are a menace to the state, (2) if they are perpetuated by heredity, and (3) if sterilization seems to be the most effective means available for dealing with them, or with certain aspects of them".[13]

What was the outcome of all this? Gosney and Popenoe found a remarkable decrease in sexual delinquency. They noted that "feebleminded" women committed to institutions were characteristically sex delinquents, being not prey to men but rather themselves aggressors, due to being "oversexed" and "feebly inhibited", while also lacking other interests. What they had found was that 9 out of 12 sterilized feebleminded women had been sexually delinquent before their commitment to an institution. However, after undergoing sterilization and being placed on parole, only 1 in 12 of those women was sexually delinquent. Gosney and Popenoe suggested that this decrease in sex delinquency was not due to sterilization but rather due to their careful placing and effective supervision while on parole, nevertheless noting:

> But it was the fact of sterilization, in many cases, that made it possible for the girl to be given this parole, with its opportunity to live a normal life in the community, to make herself self-supporting, to relieve the state of the expense of her maintenance, and to leave room in the institution for a more helpless case that required lifelong custodial care.[14]

Sterilization was the means to an end, even if it was not the main cause of that end.

The sterilization campaign against the mentally ill began in Germany with the Law on Preventing Hereditary Ill Progeny, of 1933, with sterilizations beginning on January 1, 1934. The established Hereditary Courts considered the cases of thousands of individuals, with sterilizations being particularly intensive during the first three years.[15] The German sterilization law of 1933 was remarkably similar to the "Model Eugenic Sterilization Law" proposed by Laughlin in 1922. There is a close correspondence between the respective categories of conditions. Still, the German law was more moderate than Laughlin's.[16] However, in 1934, Joseph DeJarnette, expert witness in the *Buck vs. Bell* case (and Mendelian poet, whom we met in Chapter 8), complained: "the Germans are beating us at our own game", and asked for broadening the scope of Virginia's sterilization law to resemble the German one.[17] In the same year, Leon Whitney, secretary of the American Eugenics Society whom we also met in Chapter 8, expressed his admiration for the sterilization law in Germany in a book titled *The Case for Sterilization*. He referred to the news from Germany that Hitler intended to have about 400,000 Germans, that is, about 1 in 100 of the population in his estimation, sterilized. "Whether this order is or is not directed exclusively at the Jews", Whitney wrote, "it is so grave a decision as to justify fully the recent discussion of it". He then praised that decision:

> Many far-sighted men and women in both England and America, however, have long been working earnestly toward something very like what Hitler has now made compulsory. Ridiculed, even vilified, they have fought courageously and steadily for the legalization of what they consider a constructive agency in the betterment of the race.[18]

In 1935, Marie E. Kopp, a representative of the American Committee on Maternal Health, had the opportunity to interview judges of the Hereditary Health Courts, as well as superintendents of hospitals, physicians, surgeons, psychiatrists, and social workers, during her six-month stay in Germany.[19] Her conclusion was that:

> The German sterilization law is not a hasty enactment, as some people believe. Educational work along eugenic lines goes back four decades. The first sterilization legislation was discussed before the Reichstag in 1907, about the time an American sterilization measure first became law in Indiana. Indeed the legal sterilization of

mental incompetents originated in the United States, although sterilization in the interest of public good was begun by Professor Forel in Zurich, Switzerland, some 40 years ago. The leaders in the German sterilization movement state repeatedly that their legislation was formulated only after careful study of the California experiment as reported by Mr. Gosney and Dr. Popenoe. It would have been impossible, they say, to undertake such a venture involving some one million people without drawing heavily upon previous experience elsewhere.[20]

The "California experiment" was therefore extremely valuable for the Nazis because they drew on that experience to develop their own sterilization laws. The United States had been both the pioneers and the role model.

The 1935 International Congress for Population Science took place in Berlin. Among the vice-presidents of the congress were Harry Laughlin and Clarence G. Campbell, president of the Eugenics Research Association. Laughlin could not make it to the congress, but Campbell went and gave a talk in which he said that Hitler had been able "to construct a comprehensive race policy of population development and improvement" that promised to be "epochal in racial history". This, for Campbell, set the pattern that other nations and racial groups should follow in order to refrain from falling behind "in their racial quality, in their racial accomplishment, and in their prospect of survival".[21] The *New York Times* issue of August 29, 1935, had a short article about Campbell's comments under the title: "U.S. EUGENIST HAILS NAZI RACIAL POLICY; Prof. C.G. Campbell Says Hitler Program of Development 'Promises to Be Epochal.'"[22] Campbell responded to criticisms like this in 1936, calling unfortunate the fact that "the anti-Nazi propaganda with which all countries have been flooded has gone far to obscure the correct understanding and the great importance of the German racial policy". He went on to explain that the German national policy aimed at attaining "the greater purity of racial stocks by selective endogamous mating and breeding", as well as with "the increased proportionate reproduction of the more competent eugenic stocks; and the proportionate decrease of the incompetent and undesirable dysgenic stocks".[23] His conclusion, which is at the same time an admiration for the Nazis and an alarmist message, was this:

> Thus we have the encouraging example before us of a nation that is intelligent enough to see that its first necessity is the biological one of improving in its racial quality, and thus to augment its survival values; and that has the patriotism, the resolution, and the self-discipline to make every effort toward, and to neglect no means of accomplishing, this end. Other nations cannot afford to deceive themselves as to the nature of this effort, nor to neglect the practical certainty of its success. If through such self-deception and neglect they permit a progressive deterioration in their own racial quality and survival value, and if in subsequent generations they fall behind and fail in the inescapable competition for racial survival, the present generation could only be deemed to bear a heavy responsibility for such failure.[24]

One can imagine that such strong support coming from the United States could only broaden the acceptance of the Nazi policies in Germany.

One more example of the U.S. influence is interesting. In case arguments about the importance of eugenics and racial purity were not convincing, there is always something that taxpayers will care about: money. An illustration produced in 1936 by the Reich Propaganda Office emphasized the cost of feeding a person with a hereditary disease. The illustration showed that an entire family of healthy Germans could live for one day on the same 5.50 Reichsmarks that it cost to support one ill person for the same amount of time (Figure 10.1). A similar message was conveyed by an illustration included in the "Neues Volk", meaning "New People" a monthly, widely available, illustrated magazine of the Office of Racial Policy. In 1938, it included a poster that stated (approximately): "60,000 Reichsmark. This is how much this patient with a hereditary disease costs the people's community for life. People's comrade, this is your money too". The poster showed a physically handicapped, apparently immobile, man seated, with a nurse standing behind him. The explicit message was that such disabled people would cost a lot to taxpayers; the implicit one was that something should be done about it. If it sounds familiar, it is because we already saw in Chapter 8 that Davenport had developed a similar argument to gain support for eugenics. A solution could be given by precluding them from reproduction so that the defectives would not propagate their "bad" germ.

But the Nazis took the implementation to unprecedented levels. Approximately 400,000 people were sterilized in Germany between 1933 and 1945 (about half of which had taken place until mid-1937). Another "solution" implemented after 1939 was execution, as the death of the "defectives" was financially beneficial for every

FIGURE 10.1 Propaganda illustration of 1936 showing that the cost of supporting a single person with a hereditary disease was the same for supporting a whole family of healthy Germans at the same time. Nazis defined individuals with mental, physical, or social disabilities as "hereditarily ill" and claimed that such individuals placed both a genetic and financial burden upon society and the state. U.S. Holocaust Memorial Museum, courtesy of Roland Klemig.

healthy "Volksgenossen", that is member of the "Volk". To achieve the goal of a "healthy Volkskörper", not only "genetically healthy" subjects were "bred" in the homes of the *Lebensborn* but also "hereditary ill" people were eliminated. With the Aktion T4, more than 200,000 mentally ill and disabled people from psychiatric sanatoriums and nursing homes—"life unworthy of life" in the eyes of the Nazis—were systematically murdered between 1939 and 1945.[25] The concept of "life unworthy of life" had been formulated for the first time in a 1920 essay by jurist Karl Binding and psychiatrist Alfred Hoche, titled "Permission for the Destruction of Life Unworthy of Life" (*"Die Freigabe der Vernichtung lebensunwerten Lebens"*). The essay was influential for the development of Aktion T4, even though it also contained arguments that contradicted Nazi practices. For instance, the essay lacked any mention of racial superiority or racial criteria for elimination, whereas there was an insistence on voluntary action.[26]

As legal scholar James Whitman cogently put it, Nazism was not just a nightmare, a parenthesis in history that came out of nowhere. Nor were the Nazis some demons that emerged from some underworld until they were destroyed. They followed Western government traditions, and there were continuities with what came before, particularly in the United States.[27]

MENDELIAN GENETICS AND THE NUREMBERG LAWS

Racial purity was a main concern for the racial hygienists. A main goal of the Nazis was to avoid the mixing ("Vermischung") of "Völker" and any further penetration of Jewish blood into the German "Volk". This would be done in order to avoid the perceived degeneration that had been witnessed in other societies. By prohibiting a "Mischehe" or "Mischheirat", or mixed marriage, it would be possible to avoid the outcome of sexual mixing, namely, the birth of degenerate "Mischling" children—or "mongrels". What was particularly despised was the race defilement "Rassenschande", especially when it occurred from the sexual unions of German women and "racially inferior" Jewish men. If this sounds familiar, it is because the arguments and the laws resemble the respective anti-miscegenation laws in the United States that aimed to prohibit marriages between white women and black men. In the U.S. South, it was considered that a black person was whoever had any black ancestry at all—what came to be known as the "one-drop rule". Astonishingly, the Nazis found this too extreme to apply in their case.[28] For instance, a 1934 book that was intended as a guide for teachers, as to how to present Nazi race policies to their students, stated: "Sharp social race separation of whites and blacks has shown itself to be necessary in the United States of America, even if it leads in certain cases to human hardness" providing as an example the case of a "mixed-race" child that while having predominantly white appearance was nevertheless considered as black.[29] It is interesting that even Nazi authors found the "one-drop rule" too hard to live with. But, at the same time, there was a more difficult issue for Nazis to deal with. Whereas the identification of black people was possible on the basis of skin color, even if sometimes exaggerated, it was not possible to distinguish a Jew from a non-Jew on the basis of visible features. Furthermore, intermarriage among Jews and non-Jews had been common for a long time. Therefore, something else had to be found in order to be used as a criterion. Radicals wanted to consider as Jews all those people who had at least one Jewish grandparent.

A central concern for the Nazis was to prevent the unwanted results of mendeling out ("herausmendeln"): the sudden reappearance of ancestral traits that had been hidden for several generations in a mixed population.[30] How could such a situation be avoided? There was only one certain way to avoid this in Mendelian terms: achieving purity via homozygosity. This had been made clear already in 1909 by prominent geneticists such as Bateson and Johannsen, as we saw in Chapter 8. These considerations informed the writings of racial theorists such as Hans Günther, who in his book *Racial Studies of the German People (Rassenkunde des deutschen Volkes)*, published in 1924, noted that in Europe, in general, only the Nordic and the Western people were considered beautiful, as shown in portraits and paintings. However, from a racial perspective, pure people of any race should be considered beautiful too because they exhibited all the characteristics of their race in a consistent manner. The real problem was with the mixed-race individuals, who combined traits of more than one race. Whereas each individual race was fine by itself, the crossbreeding of races resulted in contradictory combinations of individual characteristics, which did not unite but rather diverged. He then went on to consider the shape of the nose, noting that the nasal bridge, the shape of the nostrils, the nasal root, and the tip of the nose were due to independent factors and could thus be inherited independently. As a result, these might not be combined naturally in a mixed-race individual, resulting in ugliness. This provided many opportunities for meaningful Mendelian genetics studies, Günther claimed.[31]

Günther explained further that most Germans were not hybrids resulting from the crossing of two racially distinct but purebred parents; rather, they were hybrids who themselves descended from hybrids. As a result, it was not possible to predict how their children would look, and it was possible for children to exhibit features significantly different from those of their parents. The reason for this, Günther suggested, was that it was well known that individuals could inherit certain traits even from their remote ancestors. And he gave an example: when a child with a Jewish appearance was born into a non-Jewish marriage, there was no need to assume adultery; rather, one of its parents might have carried genetic traits characteristic of Jews, which were masked in their own appearance. Thus, a previously hidden genetic trait had come to the fore. He explained that there were hereditary traits that were covering (dominant) and concealed (recessive) ("man spricht von überdeckenden (dominanten) Erbanlagen und von maskierbaren (rezessiven) Merkmalen"), such as the curly hair caused by African ancestry behaved dominantly compared to the straight hair of Europeans. The problem was that the mixing would repeatedly "mendel-out" the old forms ("die alten Formen 'herausmendeln'")[32]As a result, recessive factors would be maintained and keep spreading from one generation to the next. So what could be the solution to this problem? Achieving homozygosity, or at least avoiding mixing!

You may now wonder whether there were any criticisms of such views. This was the time that Morgan first voiced his concerns about eugenics, as we saw, with Hogben, Haldane, and others to follow. However, there were virtually no biologists in Germany who expressed any such kind of concern.[33] Although Günther's work earned praise from people like Lenz and Fischer, others were less enthusiastic. A long-standing critic was the social scientist Friedrich Otto Hertz, who in 1933 was forced to resign from his university post due to his Jewish origins and moved to Vienna, where he lived until 1938 when he moved again to London. Whereas he accepted the idea that races actually existed in nature, he thought they were rarely pure and did not accept that each had specific mental features. He also rejected the idea that intelligence was entirely due to inheritance. One of his major works was *Rasse und Kultur*, published in 1925, which was translated into English and published as *Race and Civilization* in 1928.[34] Here is what he had to say at the time:

> This belief in race penetrated into the realm of science itself, and a number of organizations were charged with the task of providing the belief with a scientific foundation. At last things went so far that associations for racial eugenics were formed, members of which were only allowed to marry after a very careful inquiry into their pedigree, in order that any mixing with an inferior race might be avoided. In all this we can see how real was the contribution of racial theories to the development and the strengthening of that attitude of mind which finally brought catastrophe upon the whole German people. An infinitely exaggerated estimate of their own power, coupled with an undervaluation of that of other peoples, made them blind to the dangers of a political programme which was bound, to end in ruin.[35]

It may not come as a surprise that Goddard's book about the Kallikak family appeared in a German edition, *Die Familie Kallikak*, in November 1933, after the Nazis came to power (this was the second German edition, the first one having been published in 1914; the second edition also appeared in a special issue of the academic journal *Pedagogisches Magazin* in 1934). Karl Wilker, who made the German translation, had this to say about the impact of the Kallikak study in Germany:

> The first printing of this book aroused considerable attention. This attention often was expressed even in doubts about the genuineness of the study. How could the history of this family be true when they lived in the "land of unlimited possibilities". In the meantime research on genetic inheritance has undergone an entirely unforeseen blooming. Questions which were only cautiously touched upon by Henry Herbert Goddard at that time . . . have resulted in the law for the prevention of sick or ill offspring dated the 14th of July, 1933 [the sterilization law]. These questions then, have since become generally interesting and significant. Just how significant the problem of genetic inheritance is, perhaps no example shows so clearly as the example of the Kallikak family.[36]

Neither the criticisms that Goddard's account of "feeblemindedness" by Myerson in 1925 and others nor Goddard's own acknowledgment in 1927 that he had been wrong (Chapter 9) made any difference. The story of the Kallikak family was easy to understand and fit well with the ideology of the Nazis. Nothing else mattered. As Wilker's words imply, Goddard's account of the inheritance of "feeblemindedness" eventually paved the way for the Nazi sterilization law.

In 1933, jurist and blind-activist Rudolf Wilhelm Kraemer expressed his concerns about the possibility of sterilizing blind people in a booklet titled *Criticism of Eugenics: From the Point of View of the Person Concerned* (*Kritik der eugenik: vom standpunkt des betroffenen*). At the beginning of the booklet, Kraemer noted that whether a disease was hereditary or not could be determined only by studying the pedigree of the people concerned. He thus asked whether blindness was inherited and to what extent. He noted that it was not blindness itself that was inherited, as it could be due to various causes in different individuals; rather it was particular causes of blindness only that could be considered as a unit of inheritance. Therefore, all cases of blindness due to external causes such as injury, poisoning, or infection like gonorrhea or syphilis should be excluded. What was then left, according to Kraemer, were cases in which blindness was congenital, that is, present since birth. But even in those cases, it was not proven that all such cases were due to heredity.[37] He then turned to the available data about blind people from various sources and estimated that if all congenitally blind people had been prevented from reproducing for 60 or 70 years, there would be 174 blind people less today, which corresponded to 0.55% of the total number. He admitted that the effectiveness of the intervention could be higher, but in no way could it exceed 1%. To increase the effectiveness of such an intervention beyond that point, it would have to be performed for long periods of time, up to 500 years. But, asked Kraemer, would people have the same beliefs about heredity 500 years later?[38]

A response to Kraemer's arguments came from Otmar Freiherr von Verschuer, head of the Division of Human Heredity in the prestigious Berlin-Dahlem–based Kaiser-Wilhelm-Institut für Anthropologie, Menschliche Erblehre und Eugenik (Kaiser Wilhelm Institute for Anthropology, Human Heredity, and Eugenics) (KWIA) and, after 1942, its second director, replacing his mentor, Eugen Fischer. Verschuer maintained his position thanks to Fischer's influences, and he was himself the mentor of Josef Mengele, who became a camp doctor at Auschwitz in 1943. Mengele regularly sent to Verschuer organs and tissues from the people murdered in Auschwitz, which sustained the latter's research throughout the war, something that he persistently declined later.[39] On the very first page of his response, Verschuer quoted Kraemer's estimation that the exclusion of blind people from reproduction could at best bring about a reduction in the number of blind persons by 1%. He agreed that if this were the case, then the eugenic control of blindness would be pointless. To make a decision therefore, Verschuer continued, it was necessary to answer two questions: what was known for sure about the heredity of blindness, and how common hereditary blindness was in the population. Considering Kraemer's conceptions of heredity as being partially erroneous, he went on to provide some general knowledge about heredity. To make his point, he provided two illustrations of dominant and Mendelian inheritance, in which he explained what happened in three cases: (1) both parents were healthy; (2) one parent had the disease, and (3) both parents had the disease (Figure 10.2). According to Verschuer, there was no doubt that the Mendelian law of heredity ("Mendelsche Gesetz der Vererbung") was valid for humans, and it was thus possible to make predictions about the offspring for several normal and pathological hereditary traits. Given this, the next question was whether blindness should be considered among those hereditary traits. After describing various examples from the literature, Verschuer concluded that hereditary blindness had been confirmed with high certainty, as the inheritance of eye diseases or anomalies, which cause blindness, had been observed in hundreds of families. Therefore, the repetition of this kind of human suffering should be prevented, and the only way to do so was by preventing those people from having offspring. Therefore, there was a

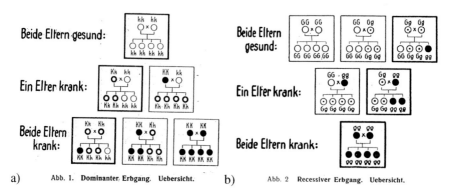

FIGURE 10.2 Overview of dominant (a) and recessive (b) inheritance provided in Verschuer's response to Kraemer. Verschuer considered necessary to explain Mendelian heredity to his readers, before developing the counterargument to Kraemer. From Verschuer (1933), reproduced with permission.

solid scientific basis for the application of practical eugenic measures, such as marriage counseling and sterilization.[40]

Like their counterparts in the United States, the German racial hygienists aimed at establishing a Mendelian pattern of inheritance for mental defects in order to justify the measures taken. If mental "defects" were due to heredity, then any environmental manipulations would be pointless for the defective individuals themselves. They thus argued that the only reasonable measures would be those preventing the propagation of these mental defects to future generations with segregation or sterilization measures. Ernst Rüdin was a key figure in this kind of research. In 1916, he wrote a paper that received international acclaim, in which he presented the method of "empirical hereditary prognosis" for nervous diseases. This was a statistically based method that differed from the previous methods that focused on the study of pedigrees. Based on a systematic study of 701 families of patients with "dementia praecox", Rüdin inferred a two-locus recessive model of Mendelian inheritance. Subsequently, he turned to the study of the inheritance of "manic-depressive insanity", which lasted between 1922 and 1925. Even though a 160-page-long manuscript of this study was produced, it was never published. It seems that Rüdin did not proceed to publication because his research failed to find a pattern of Mendelian inheritance that would be in agreement with his eugenics views.[41]

On September 15, 1935, during a Nazi Party rally held in Nuremberg, two new laws were announced, which became known as the Nuremberg Race Laws: the Reich Citizenship Law and the Law for the Protection of German Blood and German Honor. According to these laws, a person (1) *was* a Jew if they descended from at least three fully Jewish grandparents; and (2) was *counted* as a Jew, if they were a mongrel descended from two fully Jewish grandparents.[42] This seemed to provide a legal solution for identifying who is who. Racial purity was a concern for the Nazis who were committed to the idea that "every state has the right to maintain its population pure and unmixed".[43] Interestingly, a 1936 poster titled "Die Nurnberger Gesetze" ("Nuremberg Race Laws") (Figure 10.3) was used to help Germans understand the

Nuremberg Laws in a way that would resemble a textbook on Mendelian genetics. White circles represent "Aryan" Germans, black circles represent Jews, and partially shaded circles represent "mixed raced" individuals. The chart thus explained heredity in the German-bloods ("Deutschbluetiger"), the Half-breeds 2nd Grade ("Mischling 2. Grades"), the Half-breeds 1st Grade ("Mischling 1. Grades"), and the Jews ("Jude"). The Half-breeds 2nd Grade were those people who had one Jewish grandparent, whereas the Half-breeds 1st Grade were those people who had two Jewish grandparents. The chart essentially explained the various kinds of ancestry possible, which were essentially three: Germans, Jews, and mixed, and based on those the kinds of marriage that were permitted or prohibited.

It is worth noting that this choice was not accidental. As historian Amir Teicher has shown, the representations of pedigrees in the German eugenic scholarship during the 1920s and the 1930s in textbooks, articles, and legal documents had specific features that differed from those of the original pedigrees. On the one hand, important biographical/genealogical data, as well as indications about the diagnostic methods used, were removed once a pedigree was produced. Furthermore, the emphasis was put on pathological conditions and the people having them, while omitting information about healthy individuals. On the other hand, emphasis was given to the hereditary features of the individuals represented in the pedigrees, for instance, by indicating those individuals who were carriers of the respective defective allele (this was done by adding a dot within the square or circle that represented the male or female individual, respectively). In this way, pedigrees were transformed from representations of family relations to hypothetical illustrations of Mendelian inheritance.[44] Following this trend, Mendelian genetics was co-opted to help people understand the Nuremberg Laws. The chart in Figure 10.3 was astonishingly similar in terms of content (crosses) and concepts ("pure" and "mixed" individuals) to representations of Mendelian genetics that one could find in books and other essays, such as the one by Verschuer (Figure 10.2).

It is worth noting here that Hitler had praised the U.S. Immigration Act of 1924 in his book *Mein Kampf*. The Nuremberg Laws of 1935 drew strongly on the United States example. These laws subjected the Jews to a status of second-class citizenship and prohibited marriage and sexual relations between Jews and "Aryans". The similarities with the respective U.S. laws were striking. The United States was an exemplar in the creation of second-class citizenship for blacks, Chinese, Filipinos, and others, and these served as a model for the Nuremberg Laws, even if they had nothing to do with Jews. The United States also had anti-miscegenation laws, not only in the South, and these were carefully studied by the Nazi lawyers.

We thus see that in various ways Mendelian genetics had a prominent place in the public discussions and representations of hereditary disease and sterilization. Not surprisingly, there was no better place to cultivate this kind of understanding than in schools.

MENDELIAN GENETICS AND THE TEACHING OF RACIAL SCIENCE IN SCHOOLS

Here is what Hitler had to say about the importance of education of young people in 1939:

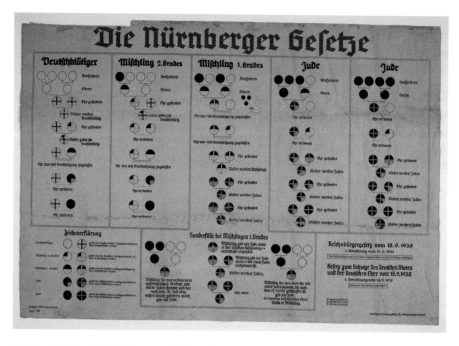

FIGURE 10.3 A 1936 chart explaining the Nuremberg Laws. It distinguishes the hierarchal difference between German-Blooded, Jews, and Mischling (part Jewish) based on their grandparent's "race". Mischling are further broken down into two grades: first class, who are half-Jewish (having two Jewish grandparents); and second class, who are one-quarter Jewish (having one Jewish grandparent). These three "races" were codified by the Reich Citizenship Law and the Law for the Protection of German Blood and German Honor, collectively known as the Nuremberg Laws, which were passed in Germany in September 1935. These laws prohibited "race defiling", that is, marriage and sexual relations between pure Germans and Jews. The chart states that Jews are not German citizens and can marry only other Jews and Mischling. The laws rejected the traditional view of Jews as members of a religious or cultural community and claimed that they were a race defined by birth and blood. This new definition included tens of thousands of people who had no Jewish cultural affiliations, such as those who had converted to Christianity. The chart was designed by Willi Hackenbergeer, the propaganda leader of the Reich Committee for Public Health Service, and it was available for purchase for one Reichsmark from the Reich Committee for Public Health Service. U.S. Holocaust Memorial Museum Collection, Gift of Virginia Ehrbar through Hillel at Kent State University.

> In my great educative work, I am beginning with the young. We older ones are used up. Yes, we are old already. We are rotten to the marrow. We have no unrestrained instincts left. We are cowardly and sentimental. We are bearing the burden of a humiliating past, and have in our blood the dull recollection of serfdom and servility. But my magnificent youngsters! Are there finer ones anywhere in the world? Look at these young men and boys! What material! With them I can make a new world.[45]

Shaping the minds of future generations was a key objective of the Nazi regime. This was attempted mainly through the schools and the youth groups, with some Nazi pedagogues considering the latter as more important than the former (by the end of this section, we see which one was actually more effective). Both were means for an education that, combined with propaganda and terror, would forge a new national identity and create a new awareness.

The most important agent in schools are teachers, and the Nazis were well aware of that. The National Socialist Teachers' League (*Nationalsozialistischer Lehrerbund*, or NSLB) had already been established in 1929, with both younger and older teachers as members. Led by Hans Schemm, who held conservative and "völkish" views, the NSLB promised changes in the public image of teachers. As soon as the Nazis came to power in 1933, the membership of NSLB grew to 12,000. Not all of those teachers were ideologically committed to the Nazi ideology though; many were concerned that they might lose their jobs or were looking for better opportunities. But gradually the membership grew even more, reaching 320,000 teachers in 1937 (97% of all teachers), while Jews and others considered unreliable by the Nazi regime were expelled. The NSLB was responsible both for assessing the political reliability of teachers and for their ideological indoctrination. The latter included all the main topics evident in the Nuremberg Laws: Nordic supremacy and the negative impact of "miscegenation" on Nordic purity, as well as information about the features of the Jewish "race". Teachers also took part in camps in order to get to know one another and instill a sense of unity among them, learn the specifics of the particular area they were staying in, and also deepen their understanding of the main points of *Mein Kampf*. Teachers left those camps with a firm acceptance of the Nazi ideology and a true belief in the Führer.[46] Not surprisingly among all those teachers, the ones teaching biology would have a key role in attaining the objectives of the Nazi regime.

The teaching of biology in schools has been considered especially important since the beginning of the Weimar Republic in 1918. The textbooks published during those years, such as Cäsar Schäffer's *Leitfaden der Biologie* (*Themes in Biology*), provided an overview of Mendelian genetics, genealogy, and even eugenics. It was also common among biology teachers to give emphasis to the teaching of heredity and eugenics in schools, something even suggested in the guidelines they received. Additional evidence for this emphasis comes from the biology exit exams, where it is clear that students had engaged with these topics. The motivation for this emphasis on heredity and eugenics in schools came primarily from the idea that biology could serve a civic function of health education. But what began with this purpose paved the way for the explicit teaching of racial science in schools after 1933.[47] After all, as Rudolph Hess put it during a mass meeting in 1934, using the popular Nazi expression also used by Schemm and first coined by Lenz in 1931, "National Socialism" was "nothing but applied biology".[48]

Sterilization and euthanasia of the "defective" could provide an answer to the concerns of the Nazis about racial purity, but these were short-term measures. For the long term, something else was necessary. Perhaps then the Nazis thought that they should carefully "educate" (meaning "indoctrinate") the future generations. It was necessary to educate people about racial purity—what it is, why it matters, and how to achieve it—and there was no better place to begin than school classrooms. Therefore, in June 1933 the Reich Interior Minister Frick circulated a plan about the teaching of biology in schools. In this it was noted not only that biology ought to be taught at least two hours per week in all school grades but also that genetics, eugenics, and racial science should be part of it. Furthermore, abundant new material was produced, such as teacher handbooks and student textbooks that focused on the teaching of genetics, eugenics, and racial science. There were many reasons why the new material was a priority, such as that there were teachers in schools who were not familiar with these topics, or that old texts that did not mention much about genetics or eugenics and certainly nothing about racial science were still available.[49]

Indeed, all textbooks began with the basics of genetics. The leading textbook of the era was the *Menschliche Erblichkeitslehre und Rassenhygiene* (*Human Heredity and Racial Hygiene*) by the geneticist Erwin Baur, along with Fischer and Lenz, which went through five editions between 1921 and 1940. This was a two-volume book that was financed, published, and promoted by the nationalist publisher J.F. Lehmanns, known for his initiative in publishing books on eugenics. Baur, Fischer, and Lenz attempted a combination of genetics, anthropology, and racial hygiene in order to cover all topics relevant to eugenics. The first volume contained chapters by Baur, Fischer, and Lenz that provided the scientific basis for the political issues of "practical racial hygiene" discussed in the second volume that was the single work of Fritz Lenz. The publication of this book in 1921 was considered an important step in the process of professionalizing racial hygiene as a scientific discipline. It also captured the general intellectual, moral, and cultural climate of the era, as Germany's defeat in World War I had created a general fear of biological decline and degeneration. Thus, it should be no surprise that an analysis of 325 contemporary reviews of the book has shown that more than 80% of the reviewers, most of whom were medical doctors focusing on its eugenics aspects, evaluated it positively, and recommended it to a variety of readers. It was the large number of positive reviews, along with the commitment of the publisher and the scientific reputation of the authors, which made this book the standard textbook on racial hygiene.[50]

The first section of the book, written by Baur, was titled "ABRISS DER ALLGEMEINEN VARIATIONS-UND ERBLICHKEITSLEHRE" ("Outline of General Variation and Heredity Theory") and contained an overview of Mendelian genetics. In the introduction, a concern was expressed about the selection processes that can rapidly lead to the deterioration of the condition of a nation. It was nevertheless realized, the story went, that such processes are becoming threatening and might lead to generation so socio-political and legislative measures had been taken. But for any such policy for all "racial hygienic endeavors (eugenics)" a scientific basis was necessary, Baur noted.[51] This basis was Mendelian genetics. Baur presented Morgan's work with *Drosophila*, as well as the results of his own experiments on *Antirrhinum majus* (the common snapdragon). Mendel was mentioned for the first time on page 25, as the person who established "alternative inheritance" (alternativen Vererbung). Baur noted the key idea of the theory of heredity on p. 34: that an apparently uniform property always depended on several independently inherited "mendelian factors". Why was this important? Because it was highly probable that the basis of a racial difference related to a Mendelian unit factor was to be found in a difference in the chemistry of chromosomes.

Another example is the 1933 book *Hereditary Science, Racial Science, Racial Hygiene* (*Erbkunde, Rassenkunde, Rassenpflege*) by Bruno Schultz, an

Austro-German anthropologist and racial hygienist, who became a prominent racial expert of the Third Reich. By 1932 he had joined both the SS and the Nazi party, subsequently holding prominent positions in the Third Reich with respect to racial science and racial policy such as being director of racial education at the SS Race and Settlement Office. His book was widely available and recommended reading. In his Foreword, Schultz wrote that his book was a comprehensive and extensive work of the kind that had not existed before, which was a necessary reference for the forthcoming teaching of hereditary biology, racial research, and racial hygiene at higher-level schools. However, he also noted that his book was not intended for school students only but also for anyone interested in understanding the principles of racial science and its applications. It is for this reason, he added, that he had included only the most essential facts and explanations while avoiding excessive details.[52] What were these essential facts? On the very first page of the book, we read about the Augustinian friar Gregor Mendel, who conducted meticulous studies on pea plants, examining their various traits, and who established the foundations for our current knowledge of inheritance in these areas.[53] The very first figure of the book is found on the next page, and it is the cross of red and white flowers of Wunderblume (*Mirabilis jalapa*, commonly known as the marvel of Peru or four o'clock flower). A few pages later we find a Punnett square and a cross between guinea pigs, where the 9:3:3:1 ratio is explained. The cross between two black and curly-haired guinea pigs gives 9 black and curly-haired, 3 black and smooth-haired black, 3 red white and curly-haired, and 1 white and smooth-haired (Figure 10.4).

One important feature of the textbooks and teaching aids of the Nazi regime was that they were the outcome of a fusion of facts from biology with aspects of Nazi ideology. A key figure was Alfred Vogel, a curriculum developer and school headmaster, who in 1938 produced a series of teaching aids that transmitted the messages that the Nazis wanted. Topics for biology education included the heredity of physical, mental, or spiritual characteristics, the heredity of disease, the heredity of physical and mental characteristics of the German race, the "law of selection", and "the Jews and the German people".[54] In one example, he drew parallels between crossbreeding in plants and racial mixing in society. By comparing "pure" flower varieties with "mixed" ones, the goal was to show how the racial purity of the German people could be maintained if they mated only with one another, or deteriorate if they mated with people from other races. To create a racially pure German nation, all those who did not belong to the Nordic race should be excluded. Therefore, to achieve this and therefore racial purity, students had to be educated about the deteriorating effects of mating with non-Nordic people (Figure 10.5).

I imagine that you will not be surprised to read that the "feebleminded" had their place too in the teachings of the Nazis's and Vogel's teaching aids. A poster titled "Feeble-mindedness in Related Families in Four Neighboring Towns" ("Schwachsinn verwandter Familien in vier Nachbarorten") was a color illustration of a landscape crossed by a river, with 4 adjacent towns alongside it (Figure 10.6). Underneath that, there was a pedigree that consisted of a couple, their four children, and their families. As ever, the squares represented males and the circles females; when these were blank, the respective individuals were healthy, whereas when they were colored this represented a disease: gold was complete "feeblemindedness", yellow was light

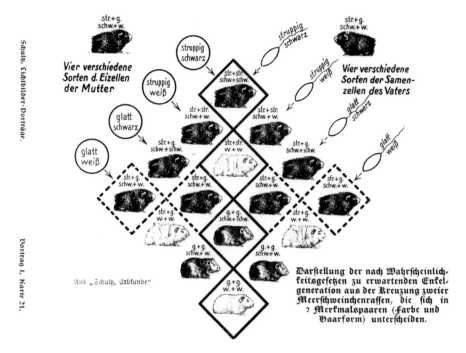

FIGURE 10.4   Mendelian genetics crossings in Bruno Schultz's 1933 book with two traits, using a typical Punnett square. This particular illustration is almost identical to the one in the book, and it has been acquired from the U.S. Holocaust Memorial Museum, courtesy of Bezirkskrankenhaus Kaufbeuren.

"feeblemindedness" and blue was alcoholism. The key feature of the chart is on the rightmost part: the marriage of two first cousins who were both "feebleminded" resulted in five children all of whom were "feebleminded" too. This is as one would expect from a disease that was a Mendelian recessive one, and the data presented in this chart stood as a confirmation of this. The chart also shows how the disease became more prevalent across generations. In the last generation we see that 10 out of 12 descendants of the initial couple have the disease.

Teaching materials of this kind were also used in classrooms, where one could find large posters with the European racial types, according to Günther's scheme, as in Figure 10.7. These were used not only for presentation but also for actual comparisons with the facial features of students. For instance, one teacher-trainee reported finding particular "fairly good distinct racial types" such as the following: "One [girl] almost pure Nordic, One almost pure Phalic, One almost pure Eastern, One not-so-pure distinct Dinaric student, and One student, that exhibited the visible markers of the East Baltic race". This was interestingly combined with the teaching of Mendelian genetics. As you can see in the upper right part of the photo, there is a typical diagram of Mendelian genetics. It is easy to imagine in this case how the teacher in the photo could combine the teaching of racial science and the teaching of

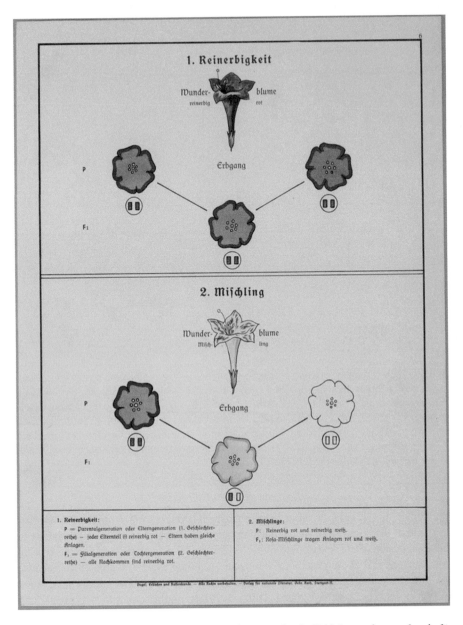

FIGURE 10.5 Poster from the 1938 racial science textbook, *Erblehre und rassenkunde für die grund-und hauptschule* (*Heredity and Race Studies for Primary and Secondary Schools*), by Alfred Vogel. The upper half depicts genetic "purity", the bottom half "mixed-breed". U.S. Holocaust Memorial Museum Collection.

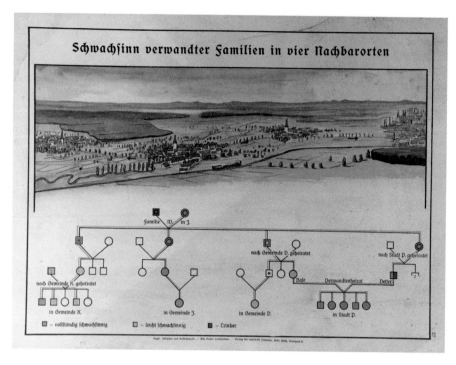

FIGURE 10.6 "Feeblemindedness" in related families in four neighboring towns. Educational poster with a family tree showing how alcoholism and feeblemindedness are passed down from a couple to their four children and their families. It is from a portfolio of teaching panels designed by Alfred Vogel in order to instruct students on the superiority of the Aryan race, the virtue of racial purity, and the burden that Jews, as well as the handicapped, mentally ill, and unfit, place on society. U.S. Holocaust Memorial Museum. Provenance: Hans Pauli. *Source Record ID*: Collections: 1990.120.1.

Mendelian genetics to talk about racial purity.[55] It is interesting to note, however, that this kind of teaching about the physical appearance of races and their comparison to the appearance of students could cause problems. The overemphasis of the Nordic traits such as fair skin, blond hair, blue eyes, and long head was dangerous because it could divide students in the same classroom between those who had those features and those who did not have them. This eventually might turn students against one another. To avoid this, the NSLB experts advised teachers to note that all "Aryan" Germans had a significant portion of Nordic blood inside them, notwithstanding their appearance. Furthermore, a suggestion made by Lenz was to focus on the mental rather than the physical traits, because it was the differences in the former that mattered most.[56]

There were even concerns about Mendel himself and his national identity that had to be addressed. Was he German? Austrian? Czechoslovakian? His name could even imply that he could have had Jewish ancestry. In 1922, Lenz argued that based on Mendel's appearance as well as on data about migrations in the area where he was

FIGURE 10.7   The teaching of racial science in a German classroom, as shown in a 1934 article. Note that right next to the chart showing racial facial types, there is another chart that is dedicated to the explication of Mendelian inheritance. *Source*: "Rassenkunde in der Volksschule", Neues Volk 7 (1934), 9. From Teicher (2020), p. 179. Courtesy of Amir Teicher.

born, there should be no doubt about his Nordic racial origins. Lenz even addressed the concerns about his name, explaining that his name in older times would have been written as "Mandel", which in the Bavarian dialect means a "small man". In 1934 a 60-page booklet on Mendelian inheritance for young readers was published. Toward its end, a girl named Magdalena became concerned that Mendel was not German. Magdalena's father confirmed that where Mendel had lived was irrelevant as he was a son of the German "Volk".[57] For the historical record, Mendel spoke, read, and wrote in German, and his nephew Alois Schindler conducted some genealogical research that concluded that about three-quarters of Mendel's ancestors were of German origin, whereas the remaining one-quarter were of Czech origin.[58]

This was the Nazi propaganda in schools throughout the biology courses. The available evidence, especially from exam tests that indicate what students did actually work on, shows that it was not necessarily the case that all students received full-blown versions of it. Rather it shows that some teachers were sensitive to the students' backgrounds or did not insist on teaching racial biology. Overall, both traditional and ideological topics were offered in biology teaching, and from the evidence we can infer that traditional topics were offered as late as 1938 and ideological topics as early as 1934. Traditional topics could be about Mendelian genetics in *Drosophila* and the work of Thomas Hunt Morgan. Ideological topics could be presented in an implicit or explicit manner. In some cases, students were taught to use Punnett squares for crosses and produce a theoretical justification for the ratios that Mendel found in his work. Whereas there were no explicit messages about human races in such cases, there were however remarks by the students about the power of heredity in humans. An astonishing example of an explicit case is that of a teacher who asked his female students to use the crosses of two *Drosophila* races in order to demonstrate the scientific foundation of one of the Nuremberg Laws, suggesting that one-quarter Jews could marry pure Aryans without any hereditary defects in the offspring.[59]

Was the effect of schools really that important? Did teaching about Mendelian genetics to justify racism make a real difference at a time when youth organizations had thousands and later millions of members? There is empirical evidence that Nazi indoctrination focusing on racial hatred was highly effective. A study with a representative sample of Germans has shown that those who grew up under the Nazi regime, between 1933 and 1945, were a lot more anti-Semitic compared to those who were born before or after that period. Not only extreme views were held by more among them, being two to three times more negative about Jews than the whole sample, but also their average views were more negative. Most interestingly, the researchers concluded that among the various possible sources of propaganda, such as the Hitler Youth, radio, printed material, film, and schooling, it was the latter that was most effective. This concerns the entire curriculum, not only the biology courses. Interestingly, school propaganda was particularly effective in those areas where people had previously held anti-Semitic beliefs.[60] Many of the boys who were secondary school students during the 1930s became the Wehrmacht soldiers who fought all over Europe and in northern Africa between 1939 and 1945. If those school boys had been convinced from their school lessons that the lives of Jews, blacks, and others were unworthy, it might have not been that hard for them to murder. If one grows up being taught that other people are a menace and of inferior quality compared to them, then it may become easier to apply the extermination plans of the regime that taught them so. This is of course no justification for what they did, but it is certainly an explanation of why so much harm was done.

Given this, what lessons can we draw for the teaching of school biology today? Of course, I am not suggesting that by teaching Mendelian genetics in schools today, there is the danger of a new Holocaust. However, letting most students who finish school have as their main knowledge Mendel's laws and Punnett squares deprives them of a deeper understanding of heredity, as most of them will not become biologists and will likely not be taught anything further. This is necessary for a deeper understanding of ourselves and our life. On the one hand, many racist arguments have been based on mistaken assumptions of biological/genetic superiority of some groups over others. As we saw, eugenics was based on this false assumption, with tragic consequences. Unfortunately, we continue to talk about genetic differences between human populations who are distinguished from one another primarily on the basis of cultural characteristics and some, few and obvious, biological characteristics, such as skin color, overlooking the fact that all humans have 99.9% of the same DNA.[61] On the other hand, the overemphasis on the potential of genes can have an important negative consequence: it relieves both the individual and society of responsibility for the manifestation of traits such as lung cancer (smoking), obesity (eating habits and patterns), and aggression (childhood abuse). It is interesting how convenient it is to attribute such traits to genes and fatalistically accept the result. By saying this, of course, I do not mean that our DNA has no role in the manifestation of such traits; obviously, it does, especially in the case of single-gene traits, which are not the norm. DNA is simply not the only factor.[62]

The question then is: what should we teach in schools if we decide not to give Mendel's peas a chance anymore?

NOTES

1 Punnett (1912), p. 137.
2 Darwin (1912), p. B2.
3 Davenport (1912a, 1912b), p. 154.
4 Weiss (2010a), pp. 27–28.
5 Weiss (2010a), p. 35; Teicher (2020), pp. 37–38, 41–43; Weindling (1989), pp. 235–238.
6 Weiss (2010a), p. 55.
7 Kühl (1994), pp. 16–18.
8 Laughlin (1914b), pp. 131–132.
9 Quoted in Weiss (2010a), p. 45.
10 Kühl (1994), pp. 19–21.
11 Gosney and Popenoe (1929) pp. 16–17.
12 Gosney and Popenoe (1929) p. ix.
13 Gosney and Popenoe (1929) p. 116.
14 Gosney and Popenoe (1929) pp. 40–41.
15 See sources in Teicher (2020), pp. 128–129.
16 Teicher (2020), pp. 141–142; Kühl (1994), pp. 38–39, 46–47.
17 Kevles (1995), p. 116; Kühl (1994), p. 45.
18 Whitney (1934), p. 7.
19 Kühl (1994), pp. 56–58.
20 Kopp (1936), p. 763.
21 Quoted in Kuhl (1994), p. 34.
22 www.nytimes.com/1935/08/29/archives/us-eugenist-hails-nazi-racial-poli-cy-prof-cg-campbell-says-hitler.html (accessed June 15, 2023).
23 Campbell (1936), p. 25.
24 Campbell (1936), p. 29.
25 Bock (1997), p. 150.
26 Brody and Cooper (2014).
27 Whitman (2017).
28 Whitman (2017), pp. 74–80.
29 Quoted in Whitman (2017), p. 127.
30 Teicher (2020), p. 92.
31 Günther (1924), pp. 241–242.
32 Günther (1924), pp. 237–239.
33 See Deichmann (1996).
34 Weindling (2010).
35 Hertz (1928/1925), p. 11.
36 Quoted in Smith (1985), pp. 161–162.

37 Kraemer (1933), pp. 7–8.
38 Kraemer (1933), p. 15.
39 Weiss (2010a, 2010b); Hansen and King (2013), pp. 158–161.
40 Verschuer (1933), pp. 5–8.
41 Kösters et al. (2015).
42 Whitman (2017), pp. 127–131.
43 Whitman (2017), p. 12.
44 Teicher (2022).
45 Rauschning (1939), pp. 246–247.
46 Pine (2010), pp. 14–17.
47 Weiss (2010a), pp. 224–229.
48 Kühl (1994), p. 36.
49 Weiss (2010a), pp. 231–232.
50 Fangerau (2005).
51 Baur, Fischer, Lenz (1927), p.XI
52 Schultz (1933). p. 3.
53 Schultz (1933). p. 5.
54 Pine (2010), pp. 43–44.
55 Teicher (2020), p. 178.
56 Weiss (2010a), p. 243.
57 Teicher (2020), pp. 145–146.
58 Orel (1996), p. 38.
59 Weiss (2010a), pp. 255–256.
60 Voigtländer and Voth (2015).
61 Kampourakis (2023).
62 Kampourakis (2021a).

Conclusions

Emending the Teaching of School Genetics

The story of eugenics shows that simplistic models about the inheritability of social traits that are not supported by solid evidence are in the best case dangerous and in the worst case fatal. This is something that we should keep in mind today in order to avoid the mistakes of the past. The application of naive and overly simplistic Mendelian models to explain all kinds of traits formed the basis for the sterilization of thousands of people all over the world, whereas in its most extreme form, it paved the way for massive exterminations. But why did these naive and oversimplistic models make so much sense to so many people, including geneticists? The central idea of these models was that there are microscopic entities inside individuals—initially called "factors", later called "genes"—that determine the traits that they have. Insofar as the respective traits seem to run across families, it seems natural to consider these factors as inheritable. In this view, some of these were dominant, others were recessive, and their combinations produced the final trait. Therefore, by carefully controlling the propagation of the "defective" factors, that is, by refraining from having individuals who carry them, it was possible to improve the genetic stock of the population and society as a whole. Or so eugenicists argued.

In light of this view, it makes sense for people to perceive these factors as the essences of what individuals are—a view that is described today as genetic essentialism. Psychologists have studied genetic essentialism in multiple studies and have arrived at the conclusion that it is a cognitive bias that emerges early in childhood and that persists into adulthood. Genetic essentialism has been defined as a tendency to infer one's characteristics and behaviors from one's perceived genetic makeup.[1] It underlies Mendelian genetics, and it is likely that this is what made eugenics possible. If people thought that internal factors (genes) were our essences, then it would make sense for them to believe that by controlling who had offspring and with whom, and who did not, they could eliminate the "defective" ones. Even though today we cannot confirm that this is actually how people thought at the time, there is some psychological evidence that this could have been the case. In an exploratory study, 1,350 participants were asked to complete various questionnaires, including one about genetic essentialism and one about genetic determinism. One of the questionnaires contained 15 questions about eugenics acceptance, where participants were asked to agree or disagree with statements such as "Sterilization of those possessing undesirable traits (e.g., a disorder) is a way to improve future generations" and "People with a criminal record should be prevented from having biological children". The

DOI: 10.1201/9781032449067-13

researchers explored possible correlations with belief in genetic essentialism or belief in genetic determinism and eugenics acceptance. Whereas the results indicated that there was limited acceptance of eugenics, a correlation of that was nevertheless found between genetic essentialism and genetic determinism. Overall, younger people and men tended to accept eugenic policies more than older people and women. Most interestingly, people who had more essentialist views of genetics tended to show more support for eugenic policies.[2]

Therefore, besides the historical evidence I have presented in Chapters 8, 9, and 10 that shows how Mendelian genetics paved the way for eugenics and racial hygiene, we also have psychological evidence about correlations between genetic essentialism and eugenic views. This is of course something that requires further research in psychology and education. However, it already gives us good reasons to reflect upon and reconsider the current school practices. By the way, I must note that what I am interested in here is what is being taught in schools. In my view, we have confused the value of the Mendelian genetics model for research with its ability to represent and explain heredity. In schools we present it as doing the latter, without explaining that its main value lies in the former. Overall, the empirical research in this domain is very limited, and therefore the conclusions that can be drawn are also limited. At the same time, there is evidence that the teaching of genetics that is based on the usual representations of genetics in textbooks can cause problems.

This has been exemplary shown by the work of Brian Donovan and his colleagues at the Biological Studies Curriculum Study. Overall, their research—based on randomized control trials—has shown that when secondary students are taught about the prevalence of monogenic disorders in different racial groups, their genetic essentialist beliefs are strengthened. In one study with eighth-grade students, the learning outcomes of two groups were compared: the students in one group had been taught about the prevalence of genetic disease, such as sickle cell anemia, in different races, whereas the students in the other group had been taught about the same diseases without any reference to race. The comparison showed that the students in the former group were more likely to believe that each race must be genetically homogeneous and discrete from others.[3] These findings were replicated with seventh- and ninth-grade students in subsequent studies.[4] In another study with eighth- and ninth-grade students, Donovan and colleagues showed that students' belief in genetic essentialism could be reduced through a five-lesson sequence about how low human genetic diversity is among groups, that is, that most genetic variation is found within rather than between continental groups. These results were replicated in two more studies with adults and 9th to 12th graders. The researchers concluded that in all three studies the intervention about human genetic variation resulted in a reduction of students' belief in genetic essentialism by changing their perception of the uniformity and the discreteness of races.[5] Finally, the researchers compared the effect of this teaching about human variation that emphasizes multifactorial genetic inheritance and human genetic diversity to the material typically taught in schools, with groups of 7th–12th-grade students. The comparison showed that the students in the former condition showed a reduction in their genetic essentialism beliefs.[6]

This is important evidence that we must change the way we teach genetics in schools. *Actually, I would suggest that we need to switch from teaching genetics—of whatever kind, including Mendelian genetics—to teaching heredity.* If you think about this, perhaps we have been doing things wrong so far and have thus lost our focus. Genetics is the scientific discipline that studies heredity, the natural phenomenon. By teaching Mendelian genetics, we have been actually teaching how science can be, and has been, done. Genetics is not about the phenomenon of heredity itself but about a particular approach to its study. However, the naive Mendelian genetics we teach in schools, which relies on a simplistic notion of genes "for" traits, which is an oversimplified representation of heredity, and the Punnett squares we use to predict the offspring, gives a distorted and misleading view of heredity. Here is then the big issue: for those who do not study biology after school, this is all they come to learn in the end. Therefore, if genetic essentialism is intuitive and naive Mendelian genetics is all that most students are taught in schools, misunderstandings that might have bad consequences and cause harm as in the past. We need to teach students in schools how heredity works, not only how we do genetics research with model systems.

Therefore, we need to go beyond the oversimplifications of heredity. In the case of the so-called monogenic traits, it is assumed that only one gene is involved and that its alleles determine one or the other version of the trait. Typical textbook examples are the shape and color of Mendel's peas, as shown in Figures 1.2 and 1.3. However, the typical description of Mendelian genetics that there exist "factors" (alleles) for yellow or green seed color or round or wrinkled seed shape is simply inaccurate. Seed color (yellow/green) is regulated by alleles *I* and *i*, with the latter retaining seed greenness not only during seed maturation but also during senescence. This suggests that allele *i* is related to the *stay-green* (*SGR*) gene, which encodes the SGR protein that is involved in the chlorophyll catabolic pathway (the exact function is not clear). Three different *i* alleles have been found; one results in the insertion of two amino acids in the SGR protein, whereas the other two result in low or no production of this protein.[7] With respect to seed shape, wrinkled seeds have higher amounts of sucrose, fructose, and glucose, resulting in higher water uptake due to osmosis. What happens is that the starch-branching enzyme SBE1, which is involved in starch synthesis, is missing in wrinkled seeds, due to an interruption of the *SBE1* gene by an insertion of 800 nucleotide pairs. Because of the lack of SBE1, there is reduced starch biosynthesis and therefore higher amounts of sucrose, fructose, and glucose.[8] Furthermore, it has been shown that the genes related to the traits that Mendel studied are not entirely independent as often assumed and that the alleles result from a range of changes at the molecular level.[9]

In addition to the characteristics that Mendel studied, another typical textbook example is the inheritance of human eye color. Textbook accounts often explain that a dominant allele *B* is responsible for brown color, whereas a recessive allele *b* is responsible for blue color—as Davenport suggested more than 100 years ago. According to such accounts, parents with brown eyes can have children with blue eyes, but it is not possible for parents with blue eyes to have children with brown eyes. The color of the eye depends on the outer tissue layer of the iris, which is called the anterior stroma. It is the density and cellular composition of this layer on which the color of the iris mostly depends. The melanocyte cells of the anterior stroma of the iris store melanin in organelles called melanosomes. When white light enters the iris, the latter can absorb or reflect a spectrum of wavelengths, giving rise to the three common iris colors (blue, green–hazel, and brown) and their variations. Blue eyes contain minimal pigment levels and melanosome numbers; green–hazel eyes have moderate pigment levels and melanosome numbers; and brown eyes are the result of high melanin levels and melanosome numbers.[10] We now know that several genes have been found to be significantly associated with eye color. It has been found that three single-nucleotide polymorphisms (SNPs) in intron 1 of the *OCA2* gene have the highest statistical association with blue eye color.[11] Other studies have shown that SNPs in the introns of gene *HERC2*, also on chromosome 15, are strongly associated with blue color. It is assumed that the variants within the *HERC2* gene are related to the expression of *OCA2* and that it is the decreased expression of the latter in iris melanocytes that is the cause of blue eye color.[12] Eye color is therefore best described as a polygenic trait.

If you keep digging further, it becomes clear that most traits are polygenic and most genes are pleiotropic, therefore gene "for" accounts is misleading. As Morgan and his colleagues explained in 1915, this expression is just a shorthand for referring to the factor or allele that, if mutated, would be the difference-maker in the occurrence of a new phenotype. But it is not the case that this alone could bring about that phenotype; it just makes a difference to the final outcome. Imagine two plants with different stem lengths: one "tall" and one "short". In theory the differences in the length of their stems could be due to genetic differences. Even if we took the extreme Mendelian model of an allele *t* that was the allele "for" the short stem of the plant (or "short") and an allele *T* that was the allele "for" the long stem of the plant (or "tall"), this would never mean that either of these two alleles would alone be responsible for the respective stem length. In contrast, what this would mean is that within a pathway with various proteins, or other molecules, being implicated in plant growth, it would be possible that allele *T* produces a protein that promotes plant growth, whereas allele *t* produces a protein that inhibits it. Therefore, *T* and *t* would be difference-makers for plant height—one would promote growth, the other would inhibit it—but none of these alone could account for the height of the plant, only for the difference from one to the other. In the same sense, it can be the change in an environmental factor that causes the difference. In fact, a difference in height between plants can be due to differences in environments. It is possible for *Arabidopsis thaliana* plants with exactly the same genotype to have different stem lengths because they have developed under different conditions. Plants exposed to mechanical stimulation by touch develop shorter petioles and bolts, which is considered an adaptive response to plant-eating animals.[13]

The question then becomes: what should school teachers teach their students? I suggest that we had better think about how we can liberate school teaching about heredity from—as most appropriately described—the "deceptive simplicity of Mendelian genetics".[14] Here is how a thoughtful geneticist put it:

> I could not imagine trying to teach genetics without starting with Mendel. Genetics is incredibly (beautifully, fascinatingly, bewilderingly) complex. . . . But to explain any of this, it helps to first understand Mendel's work and the simple experimental crosses where two versions of a gene generate forms that are sharply and certainly different, so

we start there. The important thing though is not to stop there. . . . It is easy to see how the simplified scenario that Mendel necessarily constructed for his experiments could be misunderstood as the whole picture.[15]

Well, as I mentioned earlier, this is exactly the problem. Contrary to what happens in university departments where there are whole semester courses on genetics and developmental biology, school teachers often have to stop at Mendel or Mendelian genetics (by the way, the problem is the naive Mendelian genetics, not Mendel himself). And exactly because the deterministic/essentialist view of genes "for" is easy to learn and very difficult to abandon, it is easy to stick forever if naive Mendelian genetics is all that students have been taught. Because of this, I suggest that we had better altogether get rid of the teaching of naive Mendelian genetics and Punnett squares in schools. Much as we may like it or have liked it in the past.

I think that it is not only the deceptive simplicity of Mendelian genetics but also its pedagogical attractiveness that has made it so common in school genetics teaching. There are at least two reasons for this pedagogical attractiveness of school Mendelian genetics:

1. It provides students with what seems like a complete package that brings together a "great man of science/lonely genius" story to associate with the science itself. Science thus takes a human face and becomes less abstract, when students learn how Mendel designed and performed the experiments, and then have the opportunity to perform similar crosses on paper themselves.
2. It provides grounds for problem-solving, such as those problems with the Punnett squares and the pedigree charts, which students enjoy and which teachers find easy to teach and assess. It is also one of the not-so-common instances in school biology education where students have the opportunity to do quantitative analyses and use mathematics.

What we currently teach in schools is the Mendelian genetics of Bateson and Punnett and not even their own sophisticated understanding but rather a naive form of it. I have shown that this naive understanding of Mendelian genetics provided the scientific basis for eugenics, for the thinking that there are genes "for" traits, and for naive genetic accounts of complex human behavior. Do we want to keep doing this, or do we want to instill in students the deeper understanding of Hogben and Morgan? Do we want to get rid of the naivety that characterizes popular accounts of human behavior? If we do, then we need to stop teaching naive Mendelian genetics and start teaching heredity in the context of development. Morgan and Hogben criticized eugenics because the deep understanding of development they had acquired during their experimental work had made them aware that heredity was not as simple as eugenicists argued. Heredity is not the transmission of characters from one adult form to another across generations, but the transmission of a developmental potential. To the best of my knowledge, there has been very little research on secondary students' conceptions of biological development. This is perhaps due to the fact that developmental biology is usually not taught in schools; I never really understood why it is not taught, even though development is as important for biology as evolution. Not discussing the details of development in secondary biology courses deprives students of an understanding of organisms and life. All students are familiar with developmental processes from their everyday experience. They have seen themselves and their siblings develop. They may have also made related observations on their pets and plants. However, this does not entail that they understand the respective developmental processes.

Development comprises two major phenomena: growth and differentiation. Growth occurs when the total number of body cells increases due to cell proliferation. At the same time, cells also differentiate and give rise to the various tissues and organs of the body. Both growth and differentiation are determined by local signals and interactions; there is no centralized coordination of development. The development of tissues and organs, and, eventually, the production of the adult form, does not only depend on genes or DNA but also on the exchange of signals among cells. Genes are involved in the production of proteins that are in turn involved in signal production, signal reception, and signal response. These signals consist of gradients of signaling proteins. During development, cells multiply, differentiate, and migrate to various parts of the developing organism. This happens in a coordinated manner but without any centralized coordination of development; cells simply respond to signals from their local environment. Whatever a cell does, or does not do, depends on the signals it receives from its neighboring cells. It is the local interactions among cells that drive the developmental processes. These localized processes also make the development of different organs relatively independent.[16] Development is characterized by both plasticity and robustness. Developmental robustness is the capacity of individuals of the same species to exhibit the general characteristics of their species irrespective of the environment they live in. Developmental plasticity is the capacity of individuals of the same species with the same genotype to produce different phenotypes during development as a response to local environmental conditions.[17] Developmental plasticity and robustness are two complementary aspects of development that indicate its complexity.

Perhaps contrary to what one might expect, people do not often talk about a theory of development, at least not in the way they talk about a theory of heredity or a theory of evolution. At the philosophical level, a disagreement seems to exist about whether developmental biology can offer such a theory. On the one hand, one can argue that if we consider theories as structured sets of testable explanatory and predictive hypotheses, then establishing such theories is a major goal of contemporary developmental biology; researchers try to explain and predict phenomena, give an important role to testability, and aim for unification of the observed phenomena.[18] On the other hand, there is the view that a central theory, or a family of theories, is not necessary for structuring knowledge and inquiry in science and that developmental biology is an example of this. Developmental biologists investigate phenomena such as differentiation, growth, morphogenesis, and pattern formation on the basis of multiple questions, which form an anatomical problem-structure for developmental biology.[19] Biologists seem to be divided, too. Some have argued that in contrast to evolution and heredity where we have theories and models that are widely applicable, local events and mechanisms are very important in development and produce such a heterogeneity that generalizations are difficult to make and, thus, a general theory of development is difficult to conceive.[20] The importance of local contexts is nicely illustrated by phenomena such as developmental plasticity, already mentioned earlier, in which the environment affects which one of several phenotypes encoded in DNA will emerge. However, even though developmental processes are very complex, there exist about 50–60 processes (gene regulatory networks and the process networks that they affect) that underlie development.[21] Thus, it could indeed be the case that there are some common principles of development shared by all organisms, which could form a basis for a theory.

Therefore, it is not self-evident how one could organize development education (compare this to evolution and genetics education, for instance). While considering this, we should note that in biology education we usually have in mind a rather narrow conception of development, which could be described as "adultcentric" because it focuses on multicellular organisms and on their developmental processes from a fertilized egg toward the adult form. However, this conception does not apply to all organisms; there exist organisms with multigenerational life cycles, as well as ones that consist of more than one genome. The conception of a simple, linear, developmental process does not apply in such cases. Furthermore, many unicellular organisms undergo various processes that might also be described as developmental. All this makes even the definition of what development consists of conceptually demanding.[22]

These considerations highlight some important properties of developmental processes: they are local, context-dependent, complex, and variable. If we want students in schools to understand that genes are not the essences of what and who we are and that they do not alone determine anything (perhaps with the exception of many, rare, diseases), they need to understand the developmental processes in which genes are implicated and by which traits come to be. These developmental processes are entirely neglected in the teaching of Mendelian genetics, where genes are presented to exist "for" something, without any discussion of the underlying processes. Teaching about heredity in schools makes more sense in the light of development. Of course, whether or not this will work would have to be tested empirically in classrooms.

This raises the question. Can we teach all the complexities of development to school students? Of course, not. Rather, what we have to do is to give them a sense of those complexities. To show how this could be done, I suggest that we rely on four metaphors: the "cook" metaphor, the "origami" metaphor, the "bucket" metaphor, and the "snowflake" metaphor. These metaphors could actually be used in teaching as a way of introducing students to heredity:

1. The "cook" metaphor: If two people use the same cookbook (genome) and cook a meal, the outcome (phenotype) could be very different, even though they had both followed the same recipe (DNA or genes). The expression of this information would depend on the cook (developmental system) who implemented it, because the outcome would depend on how the recipe was implemented, the experience of the cook, the type of stove used, the time available, or the conditions under which the cook worked. Consequently, it is useful to mention development alongside heredity, particularly for

multicellular organisms, as developmental processes may produce outcomes that differ from those expected by "reading" the DNA sequences alone. Simply referring to the recipe (DNA or genes) as that which determines the phenotype (meal) is wrong. But this is exactly what is currently taught in school Mendelian genetics, and this is why a discussion of developmental processes alongside genetics might help.[23]

2. The "origami" metaphor: The origami instructions provide a generative plan: they indicate when, where, and how to fold the paper in order to make a particular structure. A descriptive plan of how the origami structure would look in the end would be useless because it would provide no clues about how to generate it in the first place. In this sense, instead of thinking of the DNA of the fertilized ovum as a blueprint that contains a full description of the adult form, we can think of it as containing a set of instructions for making the adult form. This means—in the same sense as origami instructions—that these are instructions about how to make the adult form and not about how the organism will look. Cells do not implement a descriptive plan in genes but a generative one that depends on the signals they receive from their environment. It is from the combination of numerous local signals coming from the intracellular and intercellular space that cell division, proliferation, and differentiation take place during development.[24]

3. The "bucket" metaphor: Imagine two children filling a bucket of water. It would be possible to measure their relative contributions only if each one of them was filling the bucket independently because that is how we could measure how much water each one of them had added (e.g., 20 liters and 30 liters, respectively). But if one child was holding the hose over the bucket and the other turned the tap on, it would not be possible to measure how much water was due to each one of them as both contributions would be necessary in order to fill the bucket and thus could not be considered separately.[25] In this sense, the question about the relative contribution of genes and environment to the production of a character makes no sense, because they do not make independently measurable contributions.

4. The "snowflake" metaphor: The formation of snowflakes requires the simultaneous presence of two factors: temperatures below 0°C and a relative humidity sufficient for precipitation. Imagine that on a given day humidity was high at the North Pole but low at the South Pole, where the temperatures are below 0°C anyway, and that as a result there was formation of snowflakes and snowfall only in the North Pole. In this case, the variation in snowfall across the two places could be entirely accounted for by variation in relative humidity. However, this would not entail that it was relative humidity alone that caused the snowfall in the North Pole; the low temperature was also necessary for this. Therefore, accounting for variation is very different from causally explaining a phenomenon. Differences in humidity alone would be sufficient to account for differences in snowfall between the two poles only because the temperatures there are always below 0°C and not because they are unimportant in causing snow.[26]

These four metaphors convey the following messages, respectively:

1. that genes alone do not determine anything, and that the context in which their information is expressed matters ("cook" metaphor);
2. that what genes do is to provide instructions for a generative plan for development, that is, about how to make an adult form notwithstanding the final outcome ("origami" metaphor);
3. that it is not really possible to discern the contributions of genes and environments at the level of the individual, as these are interrelated and interdependent ("bucket" metaphor); and
4. that even though we may perceive a single gene as being the difference-maker for an effect, it is by no means its only cause ("snowflake" metaphor).

I therefore propose a shift in education toward designing school curricula and teaching materials that present the role of genes in their developmental contexts and for education research to investigate how students can understand this. If the aim is to educate scientifically literate citizens, then we should refrain from teaching students about genes that control or determine traits and give them instead a sense of the complexity of genomes. Mendel's work cannot provide useful lessons for today's genetics, and the oversimplistic accounts of Mendelian inheritance and the Punnett squares do not accurately represent the phenomena of heredity. This is why we had better drop Mendelian genetics altogether from the genetics curriculum: it is simply inaccurate. In doing so, we do not dishonor Mendel and his contribution. As historian and philosopher Greg Radick has nicely put it: "If we want to honour Mendel, then let us read him seriously, which is to say historically, without back-projecting the doctrinaire Mendelism that came later. Study Mendel, but let him be part of his time".[27] If we teach about Mendel, we should teach about what he did and why he did it in his own context and not present his work in anachronistic terms, as I explained in Part I of the present book. In addition, I believe that we should also dissociate his name from the naive model that provided the basis for eugenics, as I explained in Part II of the present book.

As a last word, I should note again Mendel's work was important but not for the reasons that the common hero-worshipping about him suggests. Historian Robert Olby has (once again) provided the single best account of the importance of Mendel's work:

Mendel's analysis of hybridization in terms of the independent character-pairs and their conformity with a combination series sets him apart from his contemporaries and his rediscoverers. *The treatment of hybrid variability in terms of the character-pair gave rise to the conception of two-unit hereditary determination, whereas cytological theory stimulated speculation in terms of multiple-unit hereditary determination.*[28]

What Mendel did was to develop a brilliant experimental system for the study of the transmission of character pairs across generations during hybridization, which later was used for the study of the transmission of hereditary particles. But for this conceptual shift to happen, the work that occurred in cytology between 1865 and 1900 was crucial. The conceptualization of hereditary particles in the nuclei of cells and the independence of the germ line from the rest of the body were conceptions that were fully formed toward the end of the 19th century and paved the way for the science we now call genetics. This was not Mendel's achievement, but he did indeed develop a framework that could be adapted for this kind of work in a new context. Mendel's legacy is worth celebrating but for the right reasons.

NOTES

1 Heine et al. (2019).
2 Cheung et al. (2021).
3 Donovan (2014).
4 Donovan (2016,2017).
5 Donovan et al. (2019).
6 Donovan et al. (2020).
7 Sato et al. (2007).
8 Bhattacharyya et al. (1990).
9 Reid and Ross (2011).
10 Sturm and Frudakis (2004).
11 Duffy et al. (2007); see also Frudakis et al. (2007).
12 Sulem et al. (2007); Eiberg et al. (2008); Kayser et al. (2008); Sturm et al. (2008).
13 Braam and Davis (1990); Pigliucci (2005).
14 McLysaght (2022).
15 McLysaght (2022), p. 2.
16 Davies (2014), pp. 38–40, 132, 251–252.
17 Bateson and Gluckman (2011), pp. 4–5, 8.
18 Pradeu (2014).
19 Love (2014).
20 Arthur (2014).
21 Gilbert and Bard (2014).
22 Minelli (2011), (2021).
23 Kampourakis (2021a), p. 72.
24 Wolpert (2011), p. 11.
25 Keller (2010), pp. 8–9.
26 Moore (2002), p. 41.
27 Radick (2016).
28 Olby (1985), p. 133, emphasis in the original.

References

Abbott, S., & Fairbanks, D. J. (2016). Experiments on plant hybrids by Gregor Mendel. *Genetics, 204*(2), 407–422.

Allen, G. (1975). *Life Science in the Twentieth Century*. Cambridge: Cambridge University Press.

Allen, G. E. (1978). *Thomas Hunt Morgan: The Man and His Science*. Princeton, NJ: Princeton University Press.

Allen, G. E. (2003). Mendel and modern genetics: The legacy for today. *Endeavour, 27*, 63–68.

Arthur, W. (2014). General theories of evolution and inheritance, but not development? In A. Minelli & T. Pradeu (Eds.), *Towards a Theory of Development*. Oxford: Oxford University Press, 144–154.

Audesirk, T., Audesirk, G., & Byers, B. E. (2002). *Biology: Life on Earth* (6th ed.). Hoboken, NJ: Prentice Hall.

Barkan, E. (1992). *The Retreat of Scientific Racism: Changing Concepts of Race in Britain and the United States Between the World Wars*. Cambridge: Cambridge University Press.

Barker, D. (1989). The biology of stupidity: Genetics, eugenics and mental deficiency in the inter-war years. *The British Journal for the History of Science, 22*(3), 347–375.

Bateson, P., & Gluckman, P. (2011). *Plasticity, Robustness, Development and Evolution*. Cambridge: Cambridge University Press.

Bateson, W. (1899). Hybridisation and cross-breeding as a method of scientific investigation. *Journal of the Royal Horticultural Society, 24*, 59–66.

Bateson, W. (1902). *Mendel's Principles of Heredity: A Defence*. London: Cambridge University Press.

Bateson, W. (1907). Facts limiting the theory of heredity. *Science, 26*, 649–660, 660.

Bateson, W. (1908). *The Methods and Scope of Genetics*. Cambridge: Cambridge University Press.

Bateson, W. (1909). *Mendel's Principles of Heredity*. London: Cambridge University Press.

Bateson, W. (1921–1922). Commonsense in racial problems. *Eugenics Review, 13*, 325–338.

Bateson, W., & Punnett, R. C. (1911). On the inter-relations of genetic factors. *Proceedings of the Royal Society of London. Series B, Containing Papers of a Biological Character, 84*(568), 3–8.

Bateson, W., & Saunders, E. (1901). Experimental studies in the physiology of heredity. *Reports to the Evolution Committee of the Royal Society, Report I*. London: Harrison & Sons.

Bateson, W., Saunders, E., & Punnett, R. C. (1906). Experimental studies in the physiology of heredity. *Reports to the Evolution Committee of the Royal Society, Report III*. London: Harrison & Sons.

Bateson, W., Saunders, E., & Punnett, R. C. (1908). Experimental studies in the physiology of heredity. *Reports to the Evolution Committee of the Royal Society, Report IV*. London: Harrison & Sons.

Baur, E., Fischer, E., & Lenz, F. (Eds.) (1927). *Menschliche Erblichkeitslehre und Rassenhygiene. Band I: Menschliche Erblichkeitslehre. Band II: Menschliche Auslese und Rassenhygiene (Eugenik)*. München: J.F. Lehmanns Verlag.

Bhattacharyya, M. K., Smith, A. M., Ellis, T. N., Hedley, C., & Martin, C. (1990). The wrinkled-seed character of pea described by Mendel is caused by a transposon-like insertion in a gene encoding starch-branching enzyme. *Cell, 60*(1), 115–122.

Biggs, A., Hagins, W. C., Holliday, W. G., Kapicka, C. L., & Lundgren, L. (2009). *Glencoe Science: Biology*. New York: McGraw- Hill.

Bock, G. (1997). 'Sterilization and "medical" massacres in national socialist Germany. In M. Berg & G. Cocks (Eds.), *Medicine and Modernity: Public Health and Medical Care in Nineteenth and Twentieth-Century Germany*. Cambridge: Cambridge University Press, 149–172.

Bodmer, W., Bailey, R. A., Charlesworth, B., Eyre-Walker, A., Farewell, V., Mead, A., & Senn, S. (2021). The outstanding scientist, RA Fisher: His views on eugenics and race. *Heredity, 126*(4), 565–576.

Botstein, D. (2015). *Decoding the Language of Genetics*. Cold Spring Harbor: Cold Spring Harbor Laboratory Press.

Bowler, P. J. (1989). *The Mendelian Revolution: The Emergence of Hereditarian Concepts in Modern Science and Society*. Baltimore: The Johns Hopkins University Press.

Braam, J., & Davis, R. W. (1990). Rain-, wind-, and touch- induced expression of calmodulin and calmodulin- related genes in Arabidopsis. *Cell, 60*(3), 357–364.

Brannigan, A. (1979). The reification of Mendel. *Social Studies of Science, 9*(4), 423–454.

Brannigan, A. (1981). *The Social Basis of Scientific Discoveries*. Cambridge: Cambridge University Press.

Brody, H., & Cooper, M. W. (2014). Binding and Hoche's life unworthy of life: A historical and ethical analysis. *Perspectives in Biology and Medicine, 57*(4), 500–511.

Bulmer, M. (2003). *Francis Galton: Pioneer of Heredity and Biometry*. Baltimore: Johns Hopkins University Press.

Campbell, C. G. (1936). The German racial policy. *Eugenical News, 21*(2), 25–29.

Cheung, B. Y., Schmalor, A., & Heine, S. J. (2021). The role of genetic essentialism and genetics knowledge in support for eugenics and genetically modified foods. *PLoS One, 16*(9), e0257954.

Churchill, F. B. (2015). *August Weismann: Development, Heredity, and Evolution*. Harvard, MA: Harvard University Press.

Cock, A. G., & Fordsyke, D. R. (2022). *Treasure Your Exceptions: The Science and Life of William Bateson*. Cham: Springer.

Cohen, A. (2017). *Imbeciles: The Supreme Court, American Eugenics, and the Sterilization of Carrie Buck*. New York: Penguin Books.

Corcos, A. F., & Monaghan, F. V. (1993). *Gregor Mendel's Experiments on Plant Hybrids: A Guided Study*. New Brunswick: Rutgers University Press.

Correns, C. (1900–1950). G. Mendel's law concerning the behavior of progeny of varietal hybrids. *Genetics, 35*, 33–41.

Darwin, C. (1859). *On the Origin of Species by Means of Natural Selection*. London: John Murray.

Darwin, C. (1868). *The Variation of Animals and Plants Under Domestication* (2 vols). London: John Murray.

Darwin, C. (1871). *Pangenesis. Nature, 3*, 502–503.

Darwin, C. R. (1875). *The Variation of Animals and Plants Under Domestication* (2nd ed., 2 vols.). London: John Murray.

Darwin, C. R. (1876–1882). *Recollections of the Development of My Mind & Character [Autobiography]*. CUL-DAR26.1–121. Transcribed by Kees Rookmaaker. Edited by John van Wyhe.

Darwin, L. (1912). Introduction. In *Problems in Eugenics: Papers Communicated to the First International Eugenics Congress*, The University of London, London, July 24–30, B2–B3.

Davenport, C. B. (1909). Influence of heredity on human society. *The Annals of the American Academy of Political and Social Science, 34*(1), 16–21.

Davenport, C. B. (1910). The imperfection of dominance and some of its consequences. *The American Naturalist, 44*(519), 129–135.

Davenport, C. B. (1911a). *Heredity in Relation to Eugenics*. New York: Henry Holt and Company.

Davenport, C. B. (1911b). Report of committee on Eugenics. *Journal of Heredity, 6*(1), 91–94.

Davenport, C. B. (1912a). Genetics and eugenics. In *Problems in Eugenics: Papers Communicated to the First International Eugenics Congress*, The University of London, London, July 24–30, 137–138.

Davenport, C. B. (1912b). The inheritance of physical and mental traits of man and their application to eugenics. In W. E. Castle, J. M. Coulter, C. B. Davenport, E. M. East, & W. L. Tower (Eds.), *Heredity and Eugenics: A Course of Lectures Summarizing Recent Advances in Knowledge in Variation, Heredity, and Evolution and its Relation to Plant, Animal and Human Improvement and Welfare*. Chicago: University of Chicago Press, 113–138.

Davenport, G. C., & Davenport, C. B. (1907). Heredity of eye-color in man. *Science, 26*(670), 589–592.

Davies, J. A. (2014). *Life Unfolding: How the Human Body Creates Itself*. Oxford: Oxford University Press.

Davis, A. S. (1871). The "North British review" and the origin of species. *Nature, 5*(113), 161.

de Vries, H. (1889–1910). *Intracellular Pangenesis*. Chicago: Open Court Publishing.

de Vries, H. (1900–1950). Concerning the law of segregation of hybrids. *Genetics, 35*, 30–32.

Deichmann, U. (1996). *Biologists under Hitler*. Cambridge, MA: Harvard University Press.

Dobzhansky, T. (1978). *Leslie Clarence Dunn (1893–1974): A Biographical Memoir by Theodosius Dobzhansky*. Washington, DC: National Academy of Sciences.

Donovan, B. M. (2014). Playing with fire? The impact of the hidden curriculum in school genetics on essentialist conceptions of race. *Journal of Research Science Teaching, 51*, 462–496.

Donovan, B. M. (2016). Framing the genetics curriculum for social justice: An experimental exploration of how the biology curriculum influences beliefs about racial difference. *Science Education, 100*, 586–616.

Donovan, B. M. (2017). Learned inequality: Racial labels in the biology curriculum can affect the development of racial prejudice. *Journal of Research Science Teaching, 54*, 379–411.

Donovan, B. M., Semmens, R., Keck, P., Brimhall, E., Busch, K. C., Weindling, M., Duncan, A., Stuhlsatz, M., Buck Bracey, Z., Bloom, M., et al. (2019). Towards a more humane genetics education: Learning about the social and quantitative complexities of human genetic variation research could reduce racial bias in adolescent and adult populations. *Science Education, 103*, 529–560.

Donovan, B. M., Weindling, M., & Lee, D. M. (2020). From basic to humane genomics literacy. *Science & Education, 29*, 1479–1511.

Dröscher, A. (2015). Gregor Mendel, Franz Unger, Carl Nägeli and the magic of numbers. *History of Science, 53*(4), 492–508.

Duffy, D. L., Montgomery, G. W., Chen, W., Zhao, Z. Z., Le, L., James, M. R., et al. (2007). A three–single-nucleotide polymorphism haplotype in intron 1 of *OCA2* explains most human eye-color variation. *American Journal of Human Genetics, 80*(2), 241–252.

Dunn, L. C. (1965). Mendel, his work, and his place in history. *"Commemoration of the Publication of Gregor Mendel's Pioneer Experiments in Genetics," Proceedings of the American Philosophical Society, 109*(4), 189–198.

Dunn, L. C. (1965–1991). *A Short History of Genetics: The Development of Some of the Main Lines of Thought: 1864–1939*. Ames, IA: Iowa State University Press.

East, E. M. (1912). The application of biological principles to plant breeding. In W. E. Castle, J. M. Coulter, C. B. Davenport, E. M. East, & W. L. Tower (Eds.), *Heredity and Eugenics: A Course of Lectures Summarizing Recent Advances in Knowledge in Variation, Heredity, and Evolution and Its Relation to Plant, Animal and Human Improvement and Welfare*, 113–138. Chicago: The University of Chicago Press.

East, E. M. (1917). Hidden feeble-mindedness. *Journal of Heredity, 8*, 215–217.

East, E. M. (1923). Mendel and his contemporaries. *The Scientific Monthly, 16*, 225–237.

Edwards, A. W. F. (2012). Punnett's square. *Studies in History and Philosophy of Science Part C: Studies in History and Philosophy of Biological and Biomedical Sciences, 43*(1), 219–224.

Edwards, A. W. F. (2016). Punnett's square: A postscript. *Studies in History and Philosophy of Science Part C: Studies in History and Philosophy of Biological and Biomedical Sciences, 57*, 69–70.

Eiberg, H., Troelsen, J., Nielsen, M., Mikkelsen, A., Mengel-From, J., Kjaer, K. W., & Hansen, L. (2008). Blue eye color in humans may be caused by a perfectly associated founder mutation in a regulatory element located within the *HERC2* gene inhibiting *OCA2* expression. *Human Genetics, 123*, 177–187.

Erlingsson, S. J. (2016). "Enfant terrible": Lancelot Hogben's life and work in the 1920s. *Journal of the History of Biology, 49*(3), 495–526.

Fairbanks, D. J. (2022). *Gregor Mendel: His Life and Legacy*. Buffalo, NY: Prometheus Books.

Fairbanks, D. J., & Rytting, B. (2001). Mendelian controversies: A botanical and historical review. *American Journal of Botany, 88*(5), 737–752.

Fangerau, H. M. (2005). Making eugenics a public issue: A reception study of the first German compendium on racial hygiene, 1921–1940. *Science & Technology Studies, 18*(2), 46–66.

Fisher, R. A. (1924). The elimination of mental defect. *The Eugenics Review, 16*(2), 114.

Flemming, W. (1882). *Zellsubstanz, Kern und Zelltheilung*. Leipzig: F. C. W. Vogel.

Frudakis, T., Terravainen, T., & Thomas, M. (2007). Multilocus OCA2 genotypes specify human iris colors. *Human Genetics, 122*(3–4), 311–326.

Galton, F. (1865). Hereditary talent and character. *Macmillan's Magazine, 12*, 157–166 & 318–327.

Galton, F. (1869). *Hereditary Genius: An Inquiry Into Its Laws and Consequences*. London: Macmillan.

Galton, F. (1871a). I. Experiments in Pangenesis, by breeding from rabbits of a pure variety, into whose circulation blood taken from other varieties had previously been largely transfused. *Proceedings of the Royal Society of London, 19*(123–129), 393–410.

Galton, F. (1871b) Pangenesis. *Nature, 4*, 5–6.

Galton, F. (1872). II. On blood-relationship. *Proceedings of the Royal Society of London, 20*(130–138), 394–402.

Galton, F. (1876). A theory of heredity. *The Journal of the Anthropological Institute of Great Britain and Ireland, 5*, 329–348.

Galton, F. (1883). *Inquiries Into Human Faculty and Its Development*. New York: Macmillan & Co.

Galton, F. (1889). *Natural Inheritance*. London: Macmillan.

Galton, F. (1897). The average contribution of each several ancestor to the total heritage of the offspring. *Proceedings of the Royal Society of London, 61*(369–377), 401–413.

Galton, F. (1898). A diagram of heredity. *Nature, 57*, 293–294.

Galton, F. (1904). Eugenics: Its definition, scope, and aims. *American Journal of Sociology, 10*(1), 1–25.

Galton, F. (1908). *Memories of My Life*. London: Methuen & Co.

Gilbert, S. F., & Bard, J. (2014). Formalizing theories of development: a fugue on the orderliness of change. In A. Minelli & T. Pradeu (Eds.), *Towards a Theory of Development*. Oxford: Oxford University Press, 129–143.

Gliboff, S. (1998). Evolution, revolution, and reform in Vienna: Franz Unger's ideas on descent and their post-1848 reception. *Journal of the History of Biology*, 179–209.

Gliboff, S. (1999). Gregor Mendel and the laws of evolution. *History of Science, 37*(2), 217–235.

Goddard, H. H. (1910). Heredity of feeble-mindedness. *American Breeders' Magazine, 1*(3), 165–178.

Goddard, H. H. (1912). *The Kallikak Family.* New York: Macmillan Company.

Goddard, H. H. (1914). *Feeble-mindedness: Its Causes and Consequences.* New York: Macmillan Company.

Goddard, H. H. (1927). Who is a moron? *The Scientific Monthly, 24*(1), 41–46.

Goddard, H. H. (1942). In defense of the Kallikak study. *Science, 95*(2475), 574–576.

Gosney, E. S., & Popenoe, P. (1929). *Sterilization for Human Betterment: A Summary of Results of 6,000 Operations in California, 1909–1929.* New York: MacMillan.

Green, M. (2019). Chicken heads and punnett squares: Reginald punnett and the role of visualisations in early genetics research at Cambridge, 1900–1930. In J. Nall, L. Taub & F. Willmoth (Eds.), *The Whipple Museum of the History of Science: Objects and Investigations, to Celebrate the 75th Anniversary of R. S. Whipple's Gift to the University of Cambridge.* Cambridge: Cambridge University Press, 275–290.

Griffiths, A. J. F., Gelbart, W. M., Miller, J. H., & Lewontin, R. C. (1999). *Modern Genetic Analysis.* New York: WH Freeman and Company.

Günther, H. F. (1924). *Rassenkunde des deutschen Volkes.* London: JF Lehmann.

Haldane, J. B. S. (1928). *Science and Ethics.* London: Watts and Co.

Haldane, J. B. S. (1938). *Heredity and Politics.* London: George Allen & Unwin Ltd.

Hansen, R., & King, D. (2013). *Sterilized by the State: Eugenics, Race, and the Population Scare in Twentieth-Century North America.* Cambridge: Cambridge University Press.

Harris, H. (1999). *The Birth of the Cell.* New Haven: Yale University Press.

Hartl, D. L., & Orel, V. (1992). What did Gregor Mendel think he discovered? Genetics, *131*, 245–253.

Hegreness, M., & Meselson, M. (2007). What did Sutton see?: Thirty years of confusion over the chromosomal basis of Mendelism. *Genetics, 176*(4), 1939–1944.

Heine, S. J., Cheung, B. Y., & Schmalor, A. (2019). Making sense of genetics: The problem of essentialism. *Looking for the Psychosocial Impacts of Genomic Information, Special Report, Hastings Center Report, 49*(3), S19–S26.

Henig, R. M. (2000). *The Monk in the Garden: The Lost and Found Genius of Gregor Mendel, the Father of Genetics.* Boston: Houghton Mifflin Company.

Heron, D. (1907). *A First Study of the Statistics of Insanity and the Inheritance of the Insane Diathesis.* London: Dulau and Co. https://wellcomecollection.org/works/xnyrqebe

Hertz, F. (1928). *Race and Civilization.* New York: Macmillan.

Hodge, M. J. S. (1985). Darwin as a lifelong generation theorist. In D. Kohn (Ed.), *The Darwinian Heritage.* Princeton: Princeton University Press, 207–243.

Hogben, L. (1931). *Genetic Principles in Medicine and Social Science.* London. Williams and Nordgate Ltd.

Holmes, S. J., & Loomis, H. M. (1909). The heredity of eye color and hair color in man. *Biological Bulletin, 18*(1), 50–65.

Iltis, H. (2019/1932). *Life of Mendel.* New York: Routledge.

Jenkin, F. (1867). Review of the origin of species. *The North British Review, 46*, 277–318.

Jennings, H. S. (1942). *Raymond Pearl (1879–1940): A Biographical Memoir by H.S. Jennings.* Washington, DC: National Academy of Sciences.

Johannsen, W. (1909). *Elemente der exakten Erblichkeitslehre.* Jena: Gustav Fischer.

Johannsen, W. (1911). The genotype conception of heredity. *American Naturalist, 45*, 129–159.Jones, D. F. (1944). *Edward Murray East (1879–1938): A Biographical Memoir by Donald F. Jones.* Washington, DC: National Academy of Sciences.

Kampourakis, K. (2013). Mendel and the path to Genetics: Portraying science as a social process. *Science & Education, 22*(2), 293–324.

Kampourakis, K. (2015). Myth 16: That Gregor Mendel was a lonely pioneer of genetics, being ahead of his time. In R. L. Numbers & K. Kampourakis (Eds.), *Newton's Apple and Other Myths about Science*. Cambridge, MA: Harvard University Press, 129–138.

Kampourakis, K. (2017). *Making Sense of Genes*. Cambridge: Cambridge University Press.

Kampourakis, K. (2021a). *Understanding Genes*. Cambridge: Cambridge University Press.

Kampourakis, K. (2021b). Should we give peas a chance? An argument for a mendel-free biology curriculum. In M. Haskel-Ittah & A. Yarden (Eds.), *Genetics Education: Current Challenges and Possible Solutions*. Cham: Springer.

Kampourakis, K. (2023). *Ancestry Reimagined: Dismantling the Myth of Genetic Ethnicities*. New York: Oxford University Press.

Kampourakis, K. (2024). Conclusions. In K. Kampourakis (Ed.), *Darwin Mythology: Debunking Myths, Correcting Falsehoods*. Cambridge: Cambridge University Press.

Kampourakis, K., & Roumeliotou, K. (2004). Mendel was not alone. *Biologist, 51*(1), 54–57.

Kayser, M., Liu, F., Janssens, A. C. J., Rivadeneira, F., Lao, O., van Duijn, K., et al. (2008). Three genome-wide association studies and a linkage analysis identify *HERC2* as a human iris color gene. *American Journal of Human Genetics, 82*(2), 411–423.

Keller, E. F. (2010). *The Mirage of a Space between Nature and Nurture*. Durham, NC: Duke University Press.

Kevles, D. J. (1995). *In the Name of Eugenics: Genetics and the Uses of Human Heredity*. Cambridge, MA: Harvard University Press.

Klein, J., & Klein, N. (2013). *Solitude of a Humble Genius—Gregor Johann Mendel: Volume 1*. Berlin: Springer-Verlag.

Knight, T. A. (1824). XXXII. Some remarks on the supposed influence of the pollen in cross-breeding, upon the colour of the seed-coats of plants, and the qualities of their fruits. *Philosophical Magazine Series 1, 64*(317), 191–193.

Kohler, R. E. (1994). *Lords of the Fly: Drosophila Genetics and the Experimental Life*. Chicago: University of Chicago Press.

Kopp, M. E. (1936). Legal and medical aspects of eugenic sterilization in Germany. *American Sociological Review, 1*(5), 761–770.

Kösters, G., Steinberg, H., Kirkby, K. C., & Himmerich, H. (2015). Ernst Rüdin's unpublished 1922–1925 study "inheritance of manic-depressive insanity": Genetic research findings subordinated to Eugenic ideology. *PLoS Genetics, 11*(11), e1005524.

Kraemer, R. (1933). *Kritik der eugenik: vom standpunkt des betroffenen*. Berlin: Reichsdeutscher Blindenverband.

Kühl, S. (1994). *The Nazi Connection: Eugenics, American racism, and German National Socialism*. Oxford: Oxford University Press.

Laughlin, H. H. (1914a). *Report of the Committee to Study and to Report on the Best Practical Means of Cutting off the Defective Germ–Plasm in the American Population. II. The Legal, Legislative, and Administrative Aspects of Sterilization* (Bulletin No. 10B). Cold Spring Harbor, NY: Eugenics Record Office.

Laughlin, H. H. (1914b). *Report of the Committee to Study and to Report on the Best Practical Means of Cutting Off the Defective Germ-Plasm in the American Population. I. The Scope of The Committee's Work* (Bulletin No. 10A). Cold Spring Harbor, NY: Eugenics Record Office.

Laughlin, H. H. (1923). *The Second International Exhibition of Eugenics Held September 22 to October 22, 1921*. Baltimore: Williams and Wilkins Company.

Lenay, C. (2000). Hugo De Vries: from the theory of intracellular pangenesis to the rediscovery of Mendel. *Comptes Rendus de l'Académie des Sciences-Series III-Sciences de la Vie, 323*(12), 1053–1060.

Lewis, E. B. (1978). *Alfred Henry Sturtevant (1891–1970): A Biographical Memoir by Edward B. Lewis*. Washington, DC: National Academy of Sciences.

Lippmann, W. (1922). Tests of heredity intelligence. *New Republic, 32*, 328–330.

Litchfield, H. E. (Ed.) (1915). *Emma Darwin, A Century of Family Letters, 1792–1896* (Vol. 2). London: John Murray.

Love, A. C. (2014). The erotetic organization of developmental biology. In A. Minelli & T. Pradeu (Eds.), *Towards a Theory of Development*. Oxford: Oxford University Press, 33–55.

Mader, S. S. (2004). *Biology* (8th ed.). New York: McGraw-Hill.

Maderspacher, F. (2008). Theodor Boveri and the natural experiment. *Current Biology, 18*(7), R279–R286.

Maienschein, J. (2016). Garland Allen, Thomas Hunt Morgan, and development. *Journal of the History of Biology, 49*, 587–601.

Maienschein, J., Epigenesis and Preformationism. In E. N. Zalta (Ed.), *The Stanford Encyclopedia of Philosophy (Winter 2021 Edition)*. https://plato.stanford.edu/archives/win2021/entries/epigenesis/.

Matalova, A., & Matalova, E. (2022). *Gregor Mendel—The Scientist: Based on Primary Sources 1822–1884*. Cham: Springer.

Mazumdar, P. (1992). *Eugenics, Human Genetics and Human Failings: The Eugenics Society, Its Sources and Its Critics in Britain*. New York: Routledge.

McLysaght, A. (2022). The deceptive simplicity of Mendelian genetics. *PLoS Biology, 20*(7), e3001691.

Mendel, G. (1901). Experiments on plant hybridization (translated by Charles Druery, with comment by William Bateson). *Journal of the Royal Horticultural Society, 26* (Part, I), 1–32.

Mendel, G. (1950). Gregor Mendel's letters to Carl Nägeli 1866–1873. *Genetics, 35*, 1–29.

Mendel, G. J. (2020). Experiments on plant hybrids-Versuche über Pflanzen-Hybriden: New translation with commentary. Masaryk: Masaryk University Press.

Miller, K. R., & Levine, J. S. (2010). *Miller & Levine Biology*. New York: Pearson Education.

Minelli, A. (2011). Development, an open-ended segment of life. *Biological Theory, 6*, 4–15.

Minelli, A. (2021). *Understanding Development*. Cambridge: Cambridge University Press.

Moore, D. S. (2002). *The Dependent Gene: The Fallacy of "Nature vs. Nurture."* New York: Times Books/Henry Holt and Company.

Morgan, T. H. (1909). What are 'factors' in Mendelian explanations. *American Breeders Association Reports, 5*, 365–368.

Morgan, T. H. (1910). Chromosomes and heredity. *American Naturalist, 44*, 449–498.

Morgan, T. H. (1911). Random segregation versus coupling in Mendelian inheritance. *Science, 34*(873), 384.

Morgan, T. H. (1913). Factors and unit characters in Mendelian heredity. *American Naturalist, 47*, 5–16.

Morgan, T. H. (1917). The theory of the gene. *American Naturalist, 51*, 513–544.

Morgan, T. H. (1925). *Evolution and Genetics* (2nd ed.). London: Oxford University Press.

Morgan, T. H. (1926). *The Theory of the Gene*. New Haven: Yale University Press.

Morgan, T. H., Sturtevant, A. H., Muller, H. J., & Bridges, C. B. (1915). *The Mechanism of Mendelian heredity*. New York: Henry Holt and Company.

Morris, S. W. (1994). Fleeming Jenkin and the origin of species: A reassessment. *The British Journal for the History of Science, 27*(3), 313–343.

Mukherjee, S. (2016). *The Gene: An Intimate History*. New York: Scribner.

Müller-Wille, S. (2021). Gregor Mendel and the history of heredity. In M. Dietrich, M. Borrello, & O. Harman (Eds.), *Handbook of the Historiography of Biology* (Historiography of Science, Vol. 1). Cham: Springer, 105–126.

Müller-Wille, S., & Orel, V. (2007). From Linnaean species to Mendelian factors: Elements of hybridism, 1751–1870. *Annals of Science, 64*(2), 171–215.

Müller-Wille, S., & Parolini, G. (2020). Punnett squares and hybrid crosses: How Mendelians learned their trade by the book. *British Journal for the History of Science Themes, 5,* 149–165.

Müller-Wille, S., & Rheinberger, H- J. (2012). *A Cultural History of Heredity*. Chicago: University of Chicago Press.

Myerson, A. (1917). Psychiatric family studies. *American Journal of Psychiatry, 73*(3), 355–486.

Myerson, A. (1925). *The Inheritance of Mental Diseases*. Baltimore: Williams & Wilkes.

Nägeli von, C. (1898–1884). *A Mechanico-physiological Theory of Organic Evolution*. Chicago: The Open Court Publishing Co.

Nasmyth, K. (2022). The magic and meaning of Mendel's miracle. *Nature Reviews Genetics, 23,* 447–452.

Naudin, C. (1864/1862). *Nouvelles recherches sur l'hybridité dans les végétaux*. Paris: Académie des Sciences.

Niklas, K. J., & Kutschera, U. (2015). Historical revisionism and the inheritance theories of Darwin and Weismann. *The Science of Nature, 102,* 27.

Nowicki, S. (2012). *Holt McDougal Biology*. London: Holt McDougal.

Olby, R. (1967). Franz Unger and the Wiener Kirchenzeitung: An attack on one of Mendel's teachers by the editor of a catholic newspaper. *Folia Mendeliana, ii,* 29–37.

Olby, R. (1985). *Origins of Mendelism* (2nd ed.). Chicago: The University of Chicago Press.

Olby, R. C. (1966). *Origins of Mendelism*. New York: Schocken Books.

Olby, R. C. (1979). Mendel no mendelian? *History of Science, 17,* 53–72.

Olby, R. C. (2000). Horticulture: The font for the baptism of Genetics. *Nature Reviews Genetics, 1,* 65–70.

Orel, V. (1984). *Mendel* (Past Masters series). Oxford: Oxford University Press.

Orel, V. (1996). *Mendel the First Geneticist*. Oxford: Oxford University Press.

Paleček, P. (2016). Vítězslav Orel (1926–2015): Gregor Mendel's biographer and the rehabilitation of genetics in the Communist Bloc. *History and Philosophy of the Life Sciences, 38*(3), 1–12.

Paul, D. B. (1995). *Controlling Human Heredity: 1865 to the Present*. Amherst, NY: Humanity Books.

Paul, D. B. (1998). *The Politics of Heredity: Essays on Eugenics, Biomedicine, and the Nature-Nurture Debate*. Albany: SUNY Press.

Pauly, P. J. (2000). *Biologists and the Promise of American Life: From Meriwether Lewis to Alfred Kinsey*. Princeton: Princeton University Press.

Paweletz, N. (2001). Walther Flemming: Pioneer of mitosis research. *Nature Reviews Molecular Cell Biology, 2*(1), 72–75.

Pearl, R. (1908). Breeding better men. The new science of eugenics which would elevate the race by producing higher types. *The World's Work*, January, 9818–9824.

Pearl, R. (1912a) The inheritance of fecundity. In *Problems in Eugenics: Papers Communicated to the First International Eugenics Congress Held at the University of London*, London, July 24–30, 47–57.

Pearl, R. (1912b). Genetics and Eugenics. *The Eugenics Review, 3*(4), 335.

Pearl, R. (1927). The biology of superiority. *American Mercury, 12,* 257–266.

Pearson, K. (1898). The law of ancestral inheritance. *Royal Society Proceeding, LXII,* 386–412.

Pearson, K. (1906). Walter Frank Raphael Weldon. 1860–1906. *Biometrika, 5*(1), 1–52.

Pearson, K., & Moul, M. (1925). The problem of alien immigration into Great Britain, illustrated by an examination of Russian and Polish Jewish children. *Annals of Eugenics*, *1*(1), 5–54.

Peterson, E. (2024). Myth 22: That Darwin's hatred of slavery reflected his beliefs in racial equality. In K. Kampourakis (Ed.), *Darwin Mythology: Debunking Myths, Correcting Falsehoods*. Cambridge: Cambridge University Press.

Pigliucci, M. (2005). Evolution of phenotypic plasticity: Where are we going now? *Trends in Ecology & Evolution*, *20*(9), 481–486.

Pine, L. (2010). *Education in Nazi Germany*. Oxford and New York: Berg.

Poczai, P. (2022). *Heredity Before Mendel: Festetics and the Question of Sheep's Wool in Central Europe*. New York: CRC Press.

Pradeu, T. (2014). Regenerating theories in developmental biology. In A. Minelli & T. Pradeu (Eds.), *Towards a Theory of Development*. Oxford: Oxford University Press, 15–32.

Provine, W. B. (2001). *The Origins of Theoretical Population Genetics: With a New Afterword*. Chicago: University of Chicago Press.

Punnett, R. C. (1907). *Mendelism* (2nd ed.). Cambridge: Bowes & Bowes.

Punnett, R. C. (1911). *Mendelism* (3rd ed.). London: McMillan & Co.

Punnett, R. C. (1912). Marriage laws and customs. In *Problems in Eugenics: Papers Communicated to the First International Eugenics Congress Held at the University of London*, London, July 24–30, 151–155.

Punnett, R. C. (1917). Eliminating feeble-mindedness. *Journal of Heredity*, 8, 464–465.

Punnett, R. C. (1952). William Bateson and Mendel's principles of heredity. *Notes and Records of the Royal Society of London*, *9*(2), 336–347.Radick, G. (2009). Is the theory of natural selection independent of its history? In J. Hodge & G. Radick (Eds.), *The Cambridge Companion to Darwin* (2nd ed.). Cambridge: Cambridge University Press, 147–172.

Radick, G. (2016). Teach students the biology of their time. *Nature News*, *533*(7603), 293.

Radick, G. (2023). *Disputed Inheritance: The Battle over Mendel and the Future of Biology*. Chicago: University of Chicago Press.

Rauschning, H. (Ed.) (1939). *Hitler Speaks: A Series of Political Conversations with Adolf Hitler on His Real Aims*. London: Eyre & Spottiswoode.

Raven, P. H., Johnson, G. B., Losos, J. B., Mason, K. A., & Singer, S. R. (2008). *Biology* (8th ed.). New York: McGraw-Hill.

Reece, J. B., Urry, L. A., Cain, M. L., et al. (2012). *Campbell Biology* (9th ed.). New York: Pearson Education.

Reid, J. B., & Ross, J. J. (2011). Mendel's genes: Toward a full molecular characterization. *Genetics*, *189*(1), 3–10.

Reilly, P. R. (2015). Eugenics and involuntary sterilization: 1907–2015. *Annual Review of Genomics and Human Genetics*, *16*, 351–368.

Reynolds, A. (2018). The role of cells and cell theory in evolutionary thought, ca. 1840–1872. In B. K. Hall & S. A. Moody (Eds.), *Cells in Evolutionary Biology: Translating Genotypes into Phenotypes—Past, Present, Future*. London: CRC Press, 1–24.

Rheinberger, H.-J., & Müller-Wille, S. (2016). Heredity before genetics. In S. Müller-Wille & C. Brandt (Eds.), *Heredity Explored. Between Public Domain and Experimental Science, 1850–1930*. Cambridge, MA: The MIT Press, 143–166.

Rheinberger, H.-J., & Müller-Wille, S. (2017). *The Gene: From Genetics to Postgenomics*. Chicago: University of Chicago Press.

Roberts, H. F. (1929). Plant hybridization before Mendel. In *Plant Hybridization Before Mendel*. Princeton: Princeton University Press.

Robinson, G. (1979). *A Prelude to Genetics. Theories of a Material Substance of Heredity: Darwin to Weismann*. Lawrence, KS: Coronado Press.

Roll-Hansen, N. (2014). Commentary: Wilhelm Johannsen and the problem of heredity at the turn of the 19th century. *International Journal of Epidemiology, 43*(4), 1007–1013.

Rubin, G. M., & Lewis, E. B. (2000). A brief history of Drosophila's contributions to genome research. *Science, 287*(5461), 2216–2218.

Rushton, A. R. (2014). William Bateson and the chromosome theory of heredity: A reappraisal. *The British Journal for the History of Science, 47*(1), 147–171.

Rutherford, A. (2022). *Control: The Dark History and Troubling Present of Eugenics.* London: Weidenfeld & Nicholson.

Saleeby, C. W. (1909). *Parenthood and Race Culture: An Outline of Eugenics.* New York: Moffat, Yard and Company.

Sato, Y., Morita, R., Nishimura, M., Yamaguchi, H., & Kusaba, M. (2007). Mendel's green cotyledon gene encodes a positive regulator of the chlorophyll-degrading pathway. *Proceedings of the National Academy of Sciences, 104*(35), 14169–14174.

Schultz, B. K. (1933). *Erbkunde, Rassenkunde, Rassenpflege; ein Leitfaden zum Selbststudium und fuer den Unterricht.* München: JF Lehmann Verlag.

Scott, D. H. (1891). Carl Wilhelm von Nägeli. *Nature, 44*(1146), 580–583.

Shan, Y. (2020). *Doing Integrated History and Philosophy of Science: A Case Study of the Origin of Genetics.* Cham: Springer.

Smith, J. D. (1985). *Minds Made Feeble: The Myth and Legacy of the Kallikaks.* Rockville, MD: Aspen Systems Corporation.

Spencer, H. (1864). *Principles of Biology* (Vol. 1). London: Williams and Norgate.

Stamhuis, I. H., Meijer, O. G., & Zevenhuizen, E. J. (1999). Hugo de Vries on heredity, 1889–1903: Statistics, Mendelian laws, pangenes, mutations. *Isis, 90*(2), 238–267.

Stern, A. M. (2011). From legislation to lived experience: Eugenic sterilization in California and Indiana, 1907–79. In P. A. Lombardo (Ed.), *A Century of Eugenics in America: From the Indiana Experiment to the Human Genome Era.* Blookington & Indianapolis: Indiana University Press, 95–116.

Stern, C., & Sherwood, E. R. (Eds.) (1966). *The Origin of Genetics.* San Francisco: Freeman.

Stubbe, H. (1972). *History of Genetics: From Prehistoric Times to the Rediscovery of Mendel's Laws.* Cambridge, MA: MIT Press.

Sturm, R. A., Duffy, D. L., Zhao, Z. Z., Leite, F. P. N., Stark, M. S., Hayward, N. K., et al. (2008). A single SNP in an evolutionary conserved region within intron 86 of the *HERC2* gene determines human blue-brown eye color. *American Journal of Human Genetics, 82*, 424–431.

Sturm, R. A., & Frudakis, T. N. (2004). Eye colour: Portals into pigmentation genes and ancestry. *Trends in Genetics, 20*(8), 327–332.Sturtevant, A. H. (1913). The linear arrangement of six sex- linked factors in Drosophila, as shown by their mode of association. *Journal of Experimental Zoology, 14*, 43–59.

Sturtevant, A. H. (1965–2000). *A History of Genetics.* London: CSHLP/ESP.

Subramanian, S. (2020). *A Dominant Character: The Radical Science and Restless Politics of JBS Haldane.* London: Atlantic Books.

Sulem, P., Gudbjartsson, D. F., Stacey, S. N. Helgason, A., Rafnar, T., Magnusson, K. P., et al. (2007). Genetic determinants of hair, eye and skin pigmentation in Europeans. *Nature Genetics, 39*(12), 1443–1452.

Sutton, W. S. (1902). On the morphology of the chromosome group in *Brachystola magna. Biological Bulletin, IV*(1), 24–39.

Sutton, W. S. (1903). The chromosomes in heredity. *Biological Bulletin, 4*, 231–251.

Tabery, J. (2006). Looking back on Lancelot's laughter: The Lancelot Thomas Hogben Papers, University of Birmingham, Special Collections. *The Mendel Newsletter, 15*, 10–17.

Teicher, A. (2020). *Social Mendelism: Genetics and the Politics of Race in Germany, 1900–1948.* Cambridge: Cambridge University Press.

Teicher, A. (2022). How family charts became Mendelian: The changing content of pedigrees and its impact on the consolidation of genetic theory. *History of the Human Sciences*, 09526951221107558.

Trent, J. W. (2001). 'Who shall say who is a useful person?' Abraham Myerson's opposition to the Eugenics movement. *History of Psychiatry*, *12*(45), 33–57.

Tschermak, E. (1950). Concerning artificial crossing in Pisum sativum. *Genetics*, *35*, 42–47

Tschermak, E. (1951). The rediscovery of Gregor Mendel's work: An historical retrospect. *Journal of Heredity*, *42*, 163–171.

Unger, F. (1853). *Botanical Letters to a Friend*. Translated by B. Paul. London: Samuel Highley.

van Dijk, P. J., Weissing, F. J., & Ellis, T. N. (2018). How Mendel's interest in inheritance grew out of plant improvement. *Genetics*, *210*(2), 347–355.

Verschuer, O. F. V. (1933). *Blindheit und Eugenik*. Herausgegeben vom Reichsdeutschen Blindenverband E. V. Reichsspitzenverband der deutschen Blinden. Berlin, also published as Verschuer, O. F. V. (1933). Blindheit und Eugenik. *DMW-Deutsche Medizinische Wochenschrift*, *59*(33), 1290–1292.

Vines, S. H. (1880). The works of Carl Von Nägeli. *Nature*, *23*(578), 78–80.

Voigtländer, N., & Voth, H. J. (2015). Nazi indoctrination and anti-semitic beliefs in Germany. *Proceedings of the National Academy of Sciences*, *112*(26), 7931–7936.

Walpole, B., Merson- Davies, A., & L. Dann (2011). *Biology for the IB Diploma Coursebook*. Cambridge: Cambridge University Press.

Ward, W., McGonegal, R., Tostas, P., & Damon, A. (2008). *Pearson Baccalaureate; Higher Level Biology for the IB Diploma*. Harlow: Pearson Education Limited.

Weindling, P. (1989). *Health, Race and German Politics. Between National Unification and Nazism, 1870–1945*. Cambridge: Cambridge University Press.

Weindling, P. (2010). German eugenics and the widerworld: Beyond the racial state. In A. Bashford & P. Levine (Eds.), *The Oxford Handbook of the History of Eugenics*. Oxford: Oxford University Press.

Weismann, A. (1889a–1883). On heredity. In E. Poulton et al. (Eds.), *Essays upon Heredity and Kindred Biological Problems*. Oxford. www.esp.org.

Weismann, A. (1889b–1885). The continuity of the Germ-plasm as the foundation of a theory of heredity. In E. Poulton et al. (Eds.), *Essays upon Heredity and Kindred Biological Problems*. Oxford: Oxford University Press.

Weismann, A. (1893–1892). *The Germ-plasm: A Theory of Heredity*. New York: Charles Scribner's Sons.

Weismann, A. (1904). *The Evolution Theory*. Translated by J. A. & M. R. Thomson. London: Edward Arnold.

Weiss, S. F. (2010a). *The Nazi Symbiosis: Human Genetics and Politics in the Third Reich*. Chicago: University of Chicago Press.

Weiss, S. F. (2010b). After the fall: Political whitewashing, professional posturing, and personal refashioning in the postwar career of Otmar Freiherr von Verschuer. *Isis*, *101*(4), 722–758.

Weldon, W. F. R. (1902). Mendel's laws of alternative inheritance in peas. *Biometrika*, *1*(2), 228–254.

Wells, G. P. (1978). Lancelot Thomas Hogben. *Biographical Memoirs of Fellows of the Royal Society of London*, *24*, 183–221.

Whitman, J. Q. (2017). *Hitler's American Model: The United States and the Making of Nazi Race Law*. Princeton, NJ: Princeton University Press.

Whitney, L. F. (1934). *The Case for Sterilization*. New York: Frederick A. Stokes Company

Wimsatt, W. C. (2012). The analytic geometry of genetics: Part I: The structure, function, and early evolution of Punnett squares. *Archive for History of Exact Sciences*, *66*, 359–396.

Winther, R. G. (2000). Darwin on variation and heredity. *Journal of the History of Biology*, *33*(3), 425–455.

Winther, R. G. (2001). August weismann on germ-plasm variation. *Journal of the History of Biology, 34*, 518–519.

Wolff, T. (2009). *Mendel's Theatre: Heredity, Eugenics, and Early Twentieth-Century American Drama*. Cham: Springer.

Wolpert, L. (2011). *Developmental Biology: A Very Short Introduction*. Oxford: Oxford University Press.

Wood, R. J., & Orel, V. (2001). *Genetic Prehistory in Selective Breeding: A Prelude to Mendel*. Oxford and New York: Oxford University Press.

Wunderlich, R. (1983). The scientific controversy about the origin of the embryo of phanerogams in the second quarter of the 19th century (up to 1856) and MENDEL's "Versuche über Pflanzenhybriden". In V. Orel & A. Matalova (Eds.), *GREGOR MENDEL and the Foundation of Genetics, Proceedings of the Symposium "The Past, Present and Future of Genetics" held in Kuparovice, Czechoslovakia*, August 26–28. Moravian Museum, Brno, 229–235.

Zenderland, L. (2001). *Measuring Minds: Henry Herbert Goddard and the Origins of American Intelligence Testing*. Cambridge: Cambridge University Press.

Zirkle, C. (1951). Gregor Mendel & his precursors. *Isis, 42*(2), 97–104.

Index

Note: Page numbers in *italics* indicate a figure and page numbers in **bold** indicate a table on the corresponding page.